SUPER SCIENCE A·C·T·I·V·I·T·I·E·S

Favorite Lessons from Master Teachers

Beattie • Bredt • Graeber • Lyford • Martinez
Oshita • Scotchmoor • Wight

DALE SEYMOUR PUBLICATIONS

Cover design: Rachel Gage
Illustrations: Edith Allgood, Jane Dryden (Chromatography unit)

The authors and publisher gratefully acknowledge permission to reprint excerpts from "Kool-Aid Chromatography," by Christie L. Jenkins, *Science and Children* 23, no. 7 (April 1986): 25-27. Copyright 1986 by the National Science Teachers Association, 1742 Connecticut Avenue, N.W., Washington, D.C. 20009.

This publication is based on work sponsored in part by the Office of Educational Research and Improvement (OERI), Department of Education, under contract number 400-86-0009 to Far West Laboratory for Educational Research and Development. The content of this publication does not necessarily reflect the views of OERI, the Department, any other agency of the U.S. Government, or Far West Laboratory.

Order number DS09800
ISBN 0-86651-445-7

DALE
SEYMOUR
PUBLICATIONS
P.O. BOX 10888
PALO ALTO, CA 94303

5 6 7 8 9 10 11 12 13-MA-95 94 93 92

Contents

Acknowledgments

This book is an outcome of work done by the organization Teachers' Knowledge Net in Science. As co-directors of Knowledge Net, we would like to acknowledge the people who contributed to the program in general and to *Super Science Activities* in particular. First, we acknowledge the eleven participants in the second summer institute sponsored by Knowledge Net who made this book possible: Rob Beattie; Diane Bredt; Janet Graeber; Jean Lyford; Jacinta Martinez, S.N.D.; Sam Migliaccio; Beth Napier; Steven Oshita; Judith Scotchmoor; Deborah Tucker; and Tom Wight. All undertook a series of personal and professional challenges in order to share their experience and knowledge of classroom teaching and learning.

Knowledge Net was advised by a program council of practicing science teachers. We greatly appreciate the many hours of hard work and especially the patience of our colleagues and friends Lynda Chittenden, Joan Regan, Dan Sabo, and Bob Zafran.

Knowledge Net and the summer institute benefited from the contributions of some very special educators, especially Beverly Cory, Gretchen Gilfillen, Freddy Hiebert, Becky McReynolds, Richard Ponzio, Dale Seymour, and Claire Smith. We appreciate the financial support given to the program by Far West Laboratory and the professional contributions of Rosemary De La Torre, Nikki Filby, Carrie Kojimoto, Bob Peterson, and Ann Wallgren.

Finally, we are truly indebted to Darcy Kendall for editing the manuscript.

David Dwyer
Charles Fisher
November 1987

Introduction

Using the activities in this book

Super Science Activities is a collection of practical, classroom-proven lessons from the repertoires of eight highly skilled practicing teachers. The activities provide hands-on experiences to guide the learning of middle and junior high school students in earth, life, and physical sciences. The scientific method is modeled or taught explicitly throughout the activities. Materials required to do the activities are restricted to items that are readily available in many schools, or can be made inexpensively in a reasonable amount of time.

Super Science Activities contains six units and an appendix. Each unit consists of three to five lessons that focus on content usually taught as part of middle and secondary school science curricula. The book begins with two earth science units—one on plate tectonics and one on earthquakes. The next two units focus on the life science topics of genetics and ecology. The final two units present activities in the physical science areas of electricity and chromatography. The material covered in the units is not meant to be a comprehensive or exhaustive treatment of these important areas of science. Instead, the activities are intended to supplement your regular science instruction. When integrated with activities from science textbooks and other curriculum materials, this book constitutes a practical and useful resource to help you establish a strong science program. Individual students or groups of students may also undertake projects suggested in *Super Science Activities* as a jumping-off point for further study.

The activities within each unit are presented in a common format. Each unit is structured to provide sufficient information about the topic so that even teachers without a strong science background will feel comfortable and confident teaching the lesson. All the important details you need to know about each activity, such as the materials required, objectives and goals, group size, and prerequisite knowledge, are given on the first page of each lesson. A complete lesson procedure describes each activity step-by-step, and is followed by reproducible student handouts. Enrichment activities are also suggested for your students who want to do further research. Each unit follows this general format, making it easy to find specific information and making the activities easy to use.

The appendix contains information on ways to integrate computer use in science instruction. Many science classrooms either have a computer or have access to one, and this relatively new tool can be very useful in teaching and learning about science. In keeping with the practical intent of this book, the appendix discusses the application of a number of software programs that have been successfully used in science classrooms.

How the activities were developed

Super Science Activities is one result of the work of Teachers' Knowledge Net in Science, an organization of practicing science teachers in the San Francisco Bay Area. Knowledge Net was committed to strengthening science instruction in elementary and secondary schools in three ways: by capturing exemplary instructional practices and materials for extended use, by sharing the accumulated knowledge of talented teachers, and by promoting and rewarding high standards in classroom science teaching. As one way of pursuing these goals, Knowledge Net conducted summer institutes for groups of exceptional science teachers. The summer institutes provided these teachers with an opportunity to examine the teaching profession in general and their development as science teachers in particular. By reflecting on their teaching practices and systematically interacting with colleagues, participants in the institute strove to capture prime examples of their teaching craft on paper. The activities in this volume are the result of that process.

Teaching can be a remarkably exciting and fulfilling enterprise. What teacher has not experienced joy and satisfaction at seeing a student achieve an insight and thereby create a new set of personal possibilities? As anyone who has tried it surely knows, teaching is also a complex and often frustrating activity. There do not seem to be many, if any, hard and fast rules that apply to diverse groups of learners, content areas, and learning conditions.

A vast pool of knowledge and experience is available, however, to improve classroom teaching and learning. For example, research results from the sciences, educational psychology, social organization theory, and other areas have made significant contributions to science education in recent decades. In addition to these sources of information and influence, the accumulated knowledge and practices of exemplary teachers constitute a major resource for instructional improvement. A premise underlying Knowledge Net is that this resource of teachers' knowledge and experience is both undervalued and underused. By increasing the teacher-to-teacher exchange of information about teaching practices, this undervalued resource can be put to better use. It is in this spirit of teacher-to-teacher exchange that *Super Science Activities* was created.

In presenting the activities, Knowledge Net tried to emphasize the active/constructive role of teachers as opposed to the passive/mechanical role. Teachers are often creative adapters of lessons, as well as eclectic scroungers of ideas and materials. As a result, the activities presented in this book are described in enough detail to be clearly understood, but they stop well short of scripting. The intention is to present the activities in a concrete, practical manner, but to leave plenty of room for you to personalize, adapt, or extend the material. This personalization will result in individual differences in the activities taught by users of this book, even though each person will have started with a common source. On the other hand, the reproducible student handouts that accompany each activity are provided for you to use as is. If you choose to revise or extend them, so much the better.

Knowledge Net encouraged and cultivated individual teachers' voices in teaching. In keeping with this idea, the voice of each teacher-author in this book has been preserved. The overall goal is to provide classroom-proven, open-ended activities for use in middle and junior high school science instruction.

SUPER SCIENCE A·C·T·I·V·I·T·I·E·S

Are We Drifting?
A Study of Plate Tectonics

Diane Bredt
Convent of the Sacred Heart School
San Francisco, California

All my teaching experience has been at Convent of the Sacred Heart Elementary School in San Francisco, yet I have had a variety of experiences at this K-8 private girls school. I began my career as a long-term substitute in sixth, seventh, and eighth grade science. My first experience was at age 21, standing in front of a group of eighth grade girls who were mentally "gone for the summer." The rest of that year taught me to be flexible, to laugh, and to enjoy my students.

I spent the next ten years at Convent of the Sacred Heart as a fifth and sixth grade teacher sharing the responsibilities of reading, language arts, and science, in addition to the standard school duties. During those years I developed a middle school science program that emphasized process as well as content. Ten years is a long time in one classroom, so I began looking for a new challenge. The position of seventh and eighth grade science teacher became available, and I jumped for it. In retrospect, it was the best decision I could have made—the students are great and I enjoy teaching at this grade level.

To keep these energetic, demanding students interested and excited, I needed to add to my repertoire of science "tricks" and activities. To do this I began seeking others in my position—teachers doing their "thing" in their classrooms. I made connections within the San Francisco Unified School District, began taking state-sponsored workshops, became a member of a writing team for AIMS (Activities for Integrating Math and Science), and worked with Teachers' Knowledge Net. This networking enabled me to talk shop, share ideas, and keep my excitement and knowledge growing. Teaching continues to be challenging, demanding, heartwarming, and incredibly fun.

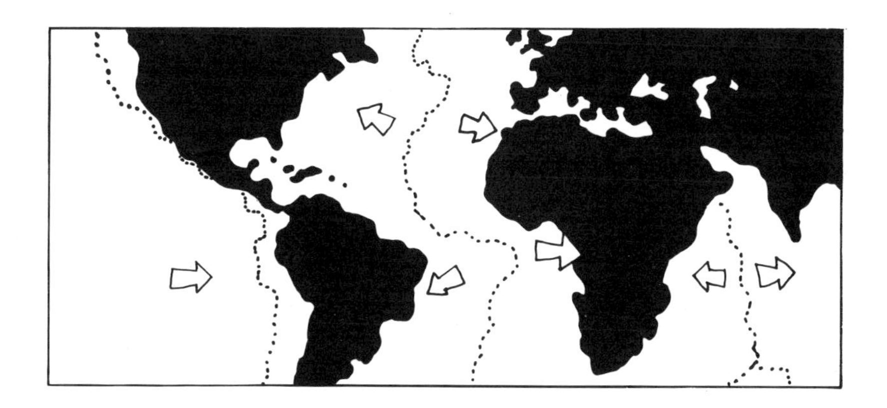

Lessons:

1. "Journey to the Center of the Earth"
2. "Do the Pieces Fit?"
3. "Where Have They Been?"
4. "Where Are the Plates?"
5. "Ride the Convection Coaster"

Overview

WHAT ARE THE FORCES that shape our planet? Why does the earth occasionally tremble beneath our feet and the sky light up with the natural fireworks of molten lava? As recently as twenty years ago there was no satisfactory answer.

This unit is an introduction to the exciting theory—plate tectonics—that attempts to answer those questions. The real hero of this revolutionary theory is Alfred Wegener, a German explorer-scientist who was obsessed with the idea that the continents were once joined together and have slowly drifted apart. Although he was not the first to speculate on the concept he called *continental drift,* he investigated it in detail and his work made other scientists consider the theory more seriously. Wegener spent his life collecting evidence to substantiate his vision, including correlating mountain ranges and fossils on different continents with matching rock formations, and correlating these index fossils on now-distant continents. Wegener died in 1930 during an expedition in Greenland, before his ideas about continental drift were accepted. The most convincing proof was to be found many years later at the bottom of the ocean, when seafloor spreading was discovered to be the mechanism that pushes the continents apart.

Although Wegener's theory was not immediately pursued, scientists continued to collect extensive data from seismic studies, maps of the ocean floors, and studies of the earth's magnetic field. By 1968 these developments led to the more encompassing theory of plate tectonics. This theory refines Wegener's concept and proposes that the earth's crust is broken into approximately 20 pieces called plates. These plates move, carrying the oceans and continents with them. The theory of plate tectonics offers an explanation for the existence of Wegener's supercontinent, Pangea, that existed 250 million years ago, as well as for the exciting geologic events that occur at plate boundaries today.

Plate tectonics is the unifying theory of geology. In addition to resolving many mysteries, this theory has enhanced our lives by enabling scientists to predict where to look for such natural resources as oil, coal, and diamonds. Earthquake zones and volcanic areas can be identified, and necessary precautions can be taken to protect people from these hazards. This revolutionary concept of plate tectonics is far from complete, but it provides scientists with a powerful foundation for analyzing the facts and data that continue to pour in.

I designed the five lessons in this unit so that students can explore plate tectonics actively. Your students will practice reading maps, researching specific information, demonstrating laboratory skills, and drawing conclusions based on available facts. After reviewing the earth's interior structure in the first lesson, your student geologists "become" Alfred Wegener as they move the continents around and determine the plate boundaries. Finally, through a laboratory activity about convection, students examine the mechanism that drives this mighty global upheaval. All these investigations will provide your students with a real sense of the global nature of plate tectonics.

Key Concepts

- ☐ The earth is not homogeneous throughout.
- ☐ The earth's interior is made of layers that differ in density and rigidity.
- ☐ These layers can be studied indirectly.
- ☐ The theory of continental drift states that the continents were once joined together and slowly drifted apart.
- ☐ The theory of plate tectonics states that the earth's crust is broken into plates that are continuously moving and have been doing so for millions of years.
- ☐ Plate tectonics explains many of the earth's topographic features:
  - shape and location of continents
  - mountain ranges
  - volcano, earthquake, and trench distribution
  - seafloor spreading
- ☐ Convection occurs when a mass of warm liquid moves to a cooler area.

Skills Used in the Lessons

- problem solving
- predicting and hypothesizing
- observing
- comparing
- diagraming
- map reading
- collecting, recording, analyzing data
- controlling, manipulating variables
- reaching conclusions

Extensions and Sources

The study of plate tectonics is a perfect introduction to the study of geologic processes. I usually follow this unit with a detailed study of earthquakes, earthquake preparedness, and volcanoes and their associated landforms. You might also focus on geologic processes specific to your particular region or location. It is very important to provide activities for each earth science concept; go outside, have your students look around for examples or evidence of the concepts they are studying. When teaching this unit, I also include a study of the geologic time scale, which links earth science and biology. I feel that studying the earth is an incomparable experience; it's exciting, violent, overwhelming, and yet it can be so peaceful. I communicate this love of the discipline to my students by continually showing my own excitement and awe. The earth is truly a dynamic place!

There are many excellent resources available for this study. The videotape, *Planet Earth Series: The Living Machine, Volume 1*, is terrific. I also use the National Geographic filmstrip set, *Geology—Our Dynamic Earth,* particularly *The Restless Earth.* I believe that earth science needs to be a very visual study and that this emphasis keeps it exciting for my students. The following books and magazines are also good sources of information.

Asimov, Isaac. *How Did We Find Out About Earthquakes?* New York: Walker & Co., 1978.

Berger, M. *Jigsaw Continents.* New York: Coward, McCann & Geoghegan, 1977.

Bolt, Bruce A. *Earthquakes: A Primer.* San Francisco: W. H. Freeman, 1978.

British Museum Geological. *Story of the Earth.* New York: Cambridge University Press, 1986.

Gore, Rick. "Our Restless Planet Earth." *National Geographic* 168, no. 2 (August 1985): 142-181.

Miller, Russell. *Continents in Collision.* Alexandria, VA: Time-Life Books, 1983.

Overbye, Dennis. "The Shape of Tomorrow." *Discover* 3, no. 11 (November 1982): 20-27.

Journey to the Center of the Earth
An exploration of the earth's interior structure

Group Size
Self-chosen groups of 2

Time Required
2-3 class periods (including optional indirect evidence activities)

Materials
- copy of Handout 1, "Journey to the Center of the Earth" on pages 12-13, and Handout 2, "The Earth's Interior" on page 14, for each student
- extra paper
- pencils
- colored pens or pencils (optional)

Key Terms

crust	outer/inner core
mantle	seismic waves
density	seismograph
epicenter	asthenosphere
Moho	lithosphere

primary waves (P-waves)
secondary waves (S-waves)

Instructional Goals
- To develop understanding that the earth's interior is composed of layers of different materials.
- To promote awareness of how scientists use indirect evidence to reach conclusions.

Student Objectives
Students will:
- Formulate hypotheses based on the available evidence.
- Determine the thickness and state of matter of each layer from the evidence given, then diagram their findings.
- Complete a chart describing the composition of each layer based on information given by the teacher.

Prerequisite Knowledge
Students should have some experience with indirect evidence activities (see Step 1 in the "Lesson Procedure" section of this lesson). Students should also know what waves are and how they move.

Advance Preparation Tme
15-30 minutes
- Review the "Background Information" and "Lesson Procedure."
- Gather materials.
- Duplicate handouts.
- Make overhead transparencies of worksheets, if desired, for class discussion and correcting.

Teacher Tips
- If possible, collect several other resources that show different views of the earth's interior. It is especially fun to have very old books for comparison.
- Start a bulletin board showing the earth's interior structure. Student drawings, diagrams, and other creative work (such as poems or written descriptions) can be added during the lesson.

Background Information

Scientists have developed a model of the interior structure of the earth and its different layers based on indirect evidence from many sources. Exactly what lies deep beneath our feet was pure speculation until the early 1900s. In 1906, after studying thousands of seismic readings using the newly invented seismograph, Andrija Mohorovicic discovered the first piece of evidence that the earth was not homogeneous. He discovered a boundary between the crust, or top layer, and the mantle directly below. This boundary was nicknamed the "Moho" in his honor. The Moho occurs about 40 km beneath the earth's surface. In 1914, Beno Gutenberg determined the thickness of the mantle by locating a second boundary, or discontinuity. However, it was not until 20 years later that the inner and outer core were discovered. Additional refinements have been made since the 1960s, differentiating the four main layers according to their densities and associated rigidity.

The outer layer of the earth, the crust, is made of igneous, sedimentary, and metamorphic rock, and averages about 40 km in depth. There are two types of crust: continental and oceanic. The continental crust ranges 15 to 70 km in thickness and is composed primarily of granite. The lower part of this granite crust rests on crystal material similar to the basaltic oceanic crust. The oceanic crust is thinner, about 8 km thick, and more dense because of its basaltic composition, which is high in iron and magnesium (see Figure 1 below).

Beneath the crust is the mantle, which is made of a different type of rock (peridotite) and extends 2900 km below the surface. The uppermost part of the mantle, to a depth of about 100 km, is made of solid rock. The crust and this uppermost part of the mantle are referred to as the *lithosphere.* Immediately below this, to a depth of about 200 km, is the plastic-like *asthenosphere.* The material in this layer flows under heat and pressure. The lower part of the mantle (200-2900 km) is the *mesosphere,* which is much more rigid. Below this, the outer core is thought to be composed of liquid iron and nickel. The very center of the earth, the inner core, is believed to be solid nickel and iron (see Figure 2 on the next page).

Our knowledge of the earth's internal structure has not been gathered directly by drilling into the crust (a 12 km drilling in 1986 is the deepest to date). Instead, indirect evidence obtained from experiments conducted on the earth's surface and in outer space has provided most of our knowledge of the earth's interior. Proof of a dense core was

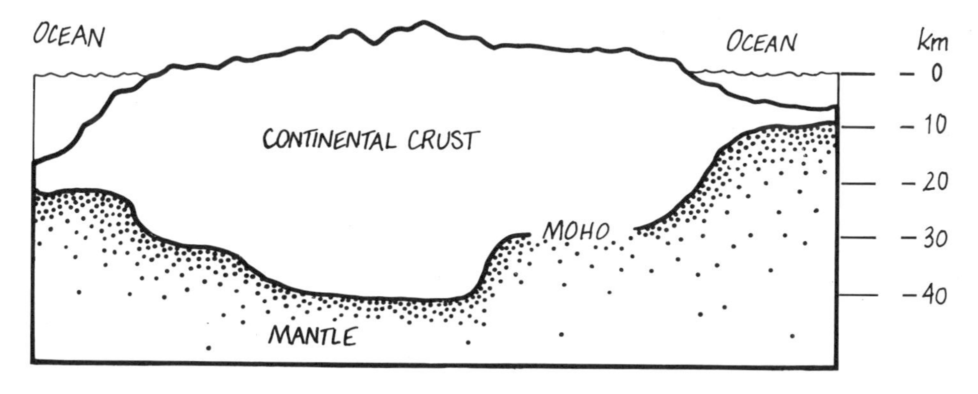

Figure 1. A cross-section of the earth's crust showing the thick continental crust and the thin oceanic crust.

obtained by determining the earth's mass and radius, which then gave the average density (5.5 g/cm) of the earth. This average density is nearly twice the density of the earth's surface rocks, which suggests a denser interior layer. The chemical analysis of meteorites has also provided clues to the internal composition of the earth, since meteorites are believed to have an origin similar to the earth's.

By studying seismic waves, specifically P-waves (primary waves) and S-waves (secondary waves), scientists have been able to identify the boundaries of the earth's various layers. P-waves and S-waves are the first two vibrations felt during an earthquake and are discussed in more detail in "A Free Ride on the Earth's Crust," on pages 42-76. The seismic waves released during an earthquake pass through the earth's interior in all directions. As these waves move through materials of different densities and temperatures, they change direction slightly due to their change in speed. Analysis of how the waves move as they pass through the earth has enabled scientists to develop a model of the earth's interior (see Figure 3 below). Continued work in seismic wave analysis with newer technologies will provide a more accurate "picture" of the earth's hidden structure.

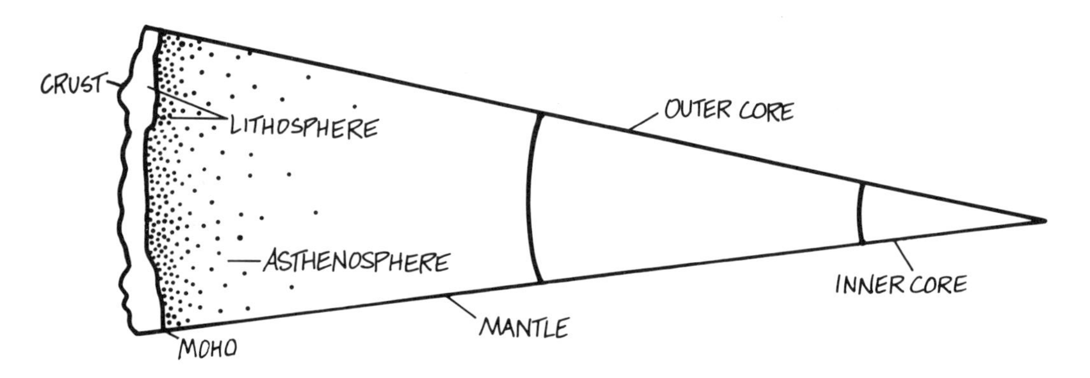

Figure 2. The earth's internal structure.

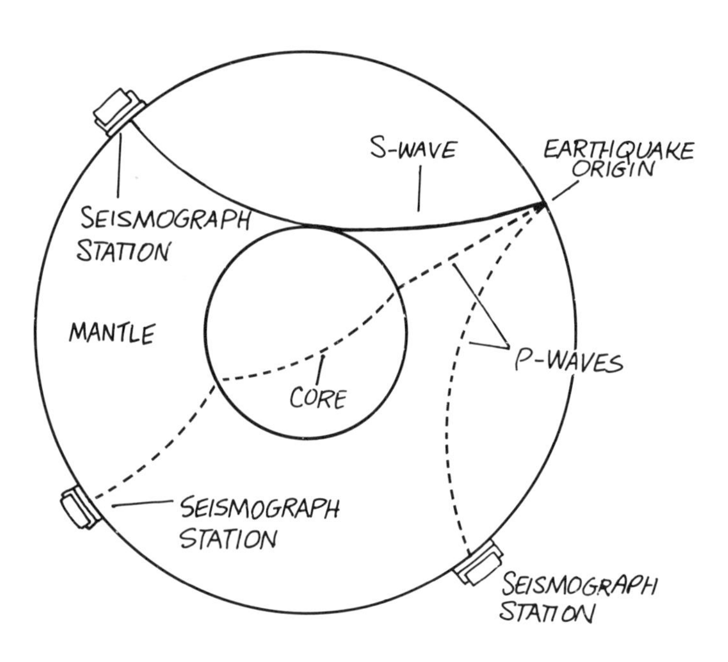

Figure 3. This diagram shows how seismic waves bend at boundaries (discontinuities) due to their change in speed. It also illustrates how scientists determined that the outer core is liquid.

Lesson Procedure

Step 1 (optional):
To produce a model of the interior of the earth, scientists must rely on indirect evidence. It is important that your students have experience and practice at gathering indirect information and drawing conclusions based on that evidence. If your students have not done activities of this nature before, consider doing one before beginning the "Journey to the Center of the Earth" activity. Two examples of these types of activities follow.

1. Have your students work in groups of two. Give one partner a small box and access to many small objects, such as pins, marbles, or buttons. One partner puts an object into the box and then gives it to the other, who then tries to determine the object's identity without looking. Students may use any of the other senses. The main objective is to show students how discoveries can be made using only indirect evidence. After working with one object, two objects can be put in the box to increase the activity's difficulty.

2. The same type of activity can be done with a piece of clay molded into a shape. Students can use "tools" such as a long pin or thin straw to help with the identification process.

Discussion of these experiences is imperative. Relate the activities to the ways scientists study the world.

Step 2:
Have your students individually draw their ideas of what the inside structure of the earth might look like, using color if they wish. Student drawings might include pictures of caves, or molten lava and underground volcanoes. Share and discuss their ideas, avoiding any mention of the "right" structure.

Step 3:
Distribute copies of Handout 1, "Journey to the Center of the Earth" on pages 12-13 (be sure to have copies of both pages). Your students should work in self-chosen groups ("geology teams") of two.

Go over the facts and data presented on the first page of the handout with your class, explaining any unfamiliar terms (such as density). Emphasize that the earthquake waves change direction when they travel through one type of rock to another.

Assure your students that the activity can be successfully solved. Help them get started by drawing a large circle on the board. Place an earthquake focus (the point at which the energy is released) on the outside edge of the circle. Explain that when an earthquake occurs it releases seismic waves called primary waves and secondary waves, or P-waves and S-waves for short. These waves are different and act differently in solids than they do in liquids.

Ask your students what happens to the seismic waves about 40 km below the surface. (They change direction.) Why? (See Fact 2 from the first page of Handout 1.) The waves encounter rock of a different density, therefore it is probably a different layer. This layer must be solid because both the P-waves and S-waves pass through it.

Have your students continue to deduce each of the earth's interior layers in the same way. Provide as much help as they need. If necessary, walk them through the next layer also. It is important that students see *how* the boundaries and composition of the layers were determined, not just that they exist.

The indirect evidence used by scientists to determine the earth's interior structure is quite important. Encourage discussion about how this information is obtained. Emphasize the excitement of problem solving.

Step 4:
Using the second page of Handout 1, ask your students to draw a diagram of the inside structure of the earth based on the given facts. Teams can use one partner's

circle as a trial worksheet and then do a good copy on the other partner's worksheet. Teams should then fill in the chart at the bottom of the worksheet.

Their diagram and chart should look something like those shown in Figure 4 below.

Step 5:
Have the teams compare their results. Ask each team to describe how the thickness of each layer was determined. Discuss any differences in results between the teams. After this discussion, compare the results with a diagram of the earth's interior in a reference book. This information could be put on the board or overhead projector.

Step 6:
Distribute copies of Handout 2, "The Earth's Interior." Discuss the material presented in "Background Information" with your class, naming the earth's layers and describing their composition. Emphasize that indirect evidence was used to determine this information. Make sure you have gone over each of the "Key Terms" listed at the beginning of this lesson. Students can complete Handout 2 as you present the information; alternatively, you might have students do additional research and then complete the handout. Or you might use Handout 2 as a transparency, completing it as a class activity on an overhead projector.

If you had your students do imaginary drawings of the earth's interior at the beginning of the lesson (refer to Step 2), you may want to post their drawings next to a large poster of the known structure of the earth's interior made by you or a student. This comparison shows students that, just as in the scientific method, an original hypothesis can be changed if the evidence indicates another conclusion is more appropriate.

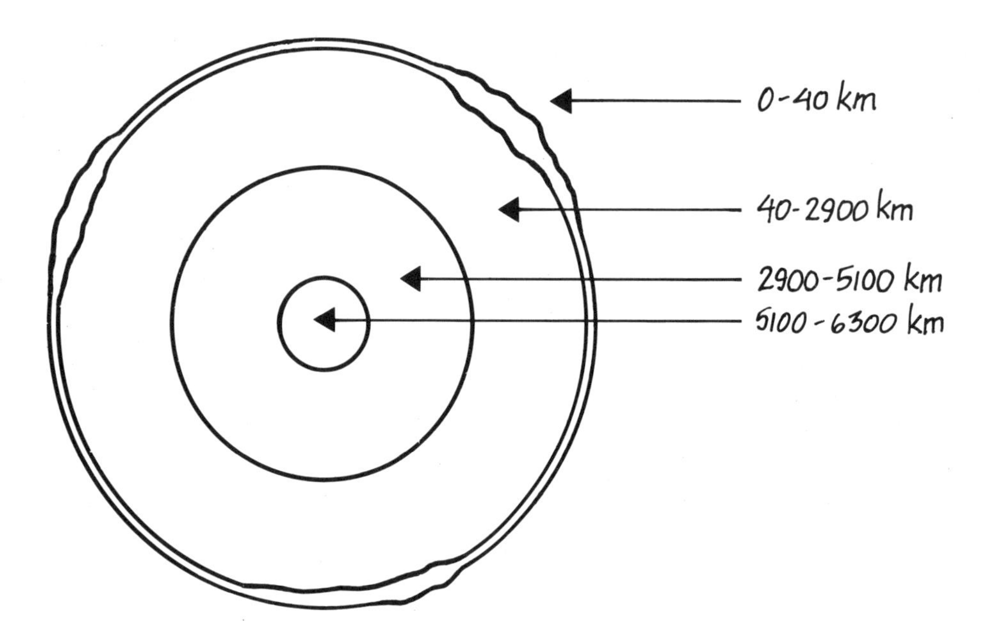

Layer	Depth	Thickness	State (solid or liquid)
#1 crust	0-40 km	40 km	solid
#2 mantle	40-2900 km	2860 km	solid
#3 outer core	2900-5100 km	2200 km	liquid
#4 inner core	5100-6300 km	1200 km	cannot be determined with information given

Figure 4. Sample answers for Handout 1.

Enrichment Activities

- ☐ After completing the second handout, students can draw their diagrams of the earth's interior structure to scale.
- ☐ Students can make 3-dimensional models of the interior structure of the earth.
- ☐ Many writing assignments can be generated from this activity, including descriptive paragraphs, fictional stories, and poetry.
- ☐ Students can trace the history of how scientists discovered these different layers, including biographical information about the people involved.
- ☐ For fun, show the movie *Journey to the Center of the Earth.* Compare the movie's version of the earth's interior to the interior structure as it has been discovered by scientists.
- ☐ Students can do research on:
  - Andrija Mohorovicic
  - the Mohole Project
  - ocean floor exploration
  - earthquake waves
  - Beno Gutenberg
  - the interior structure of other planetary bodies

Journey to the Center of the Earth

Your geologic team has spent years researching and gathering the facts and data that follow. Review this information and then draw conclusions about the earth's interior.

FACTS:

1. Earthquakes release energy that travels in waves.
2. Earthquake waves *change direction* when they pass through different types of rocks (with different densities).
3. There are two kinds of waves that travel through the earth's interior:
   - P-waves (primary waves; push-pull motion)—travel through solids and liquids.
   - S-waves (secondary waves; sideways motion)—travel through solids only.

DATA:

1. P-waves and S-waves change direction 40 km below the surface.
2. P-waves change direction abruptly about 2900 km below the surface.
3. S-waves disappear at 2900 km below the surface.
4. P-waves change direction again at 5100 km below the surface.
5. The diameter of the earth is 12,600 km.

YOUR CONCLUSIONS:

On the second page of this handout, draw a diagram of the interior of the earth based on the above facts and data. (Hint: When the waves change direction, what have they encountered?)

Answer these questions, then complete the diagram and chart.
How thick is each layer? Is it solid? Is it liquid?

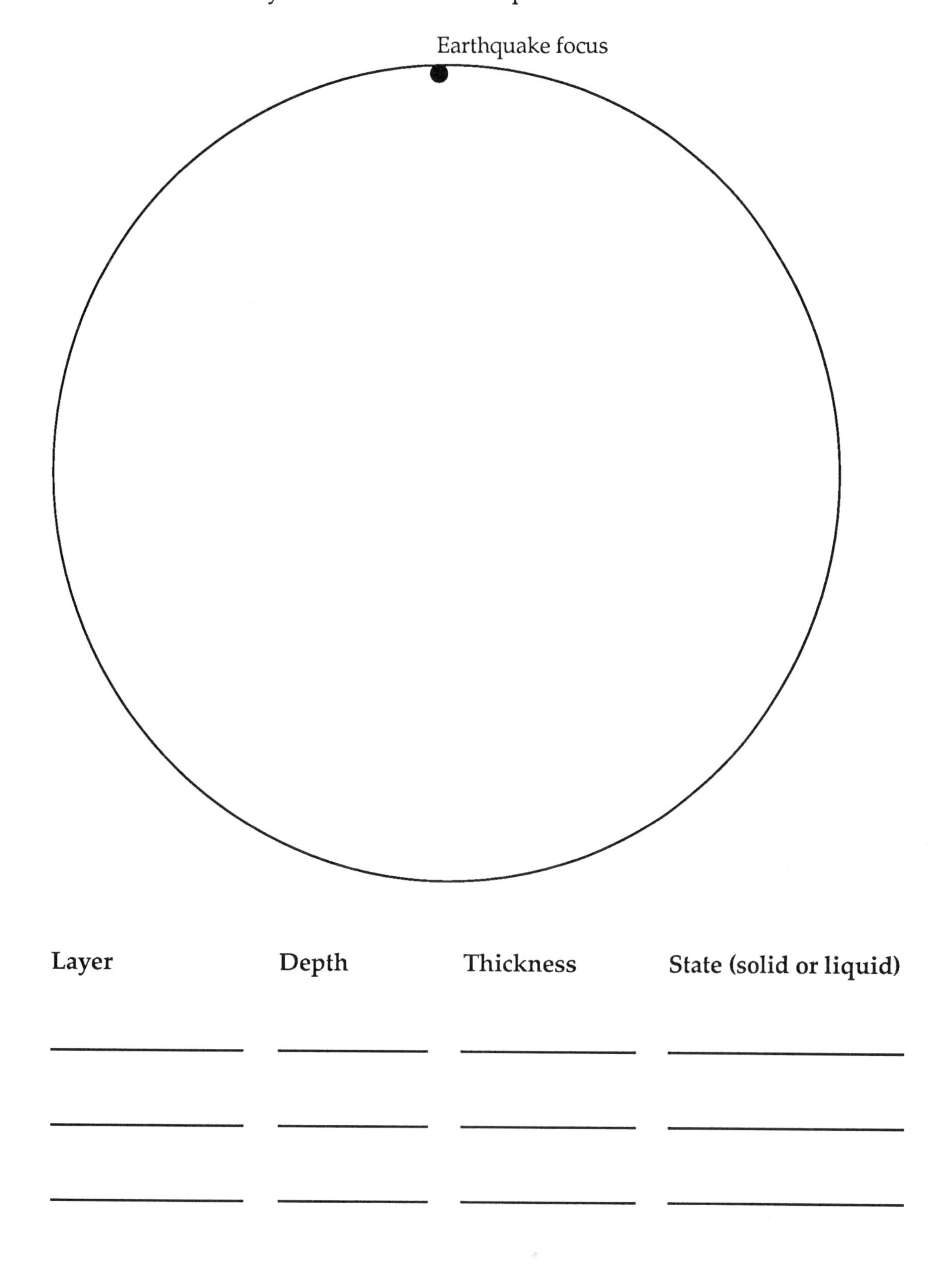

Layer	**Depth**	**Thickness**	**State (solid or liquid)**

The Earth's Interior

Label the earth's layers and the indicated boundary.

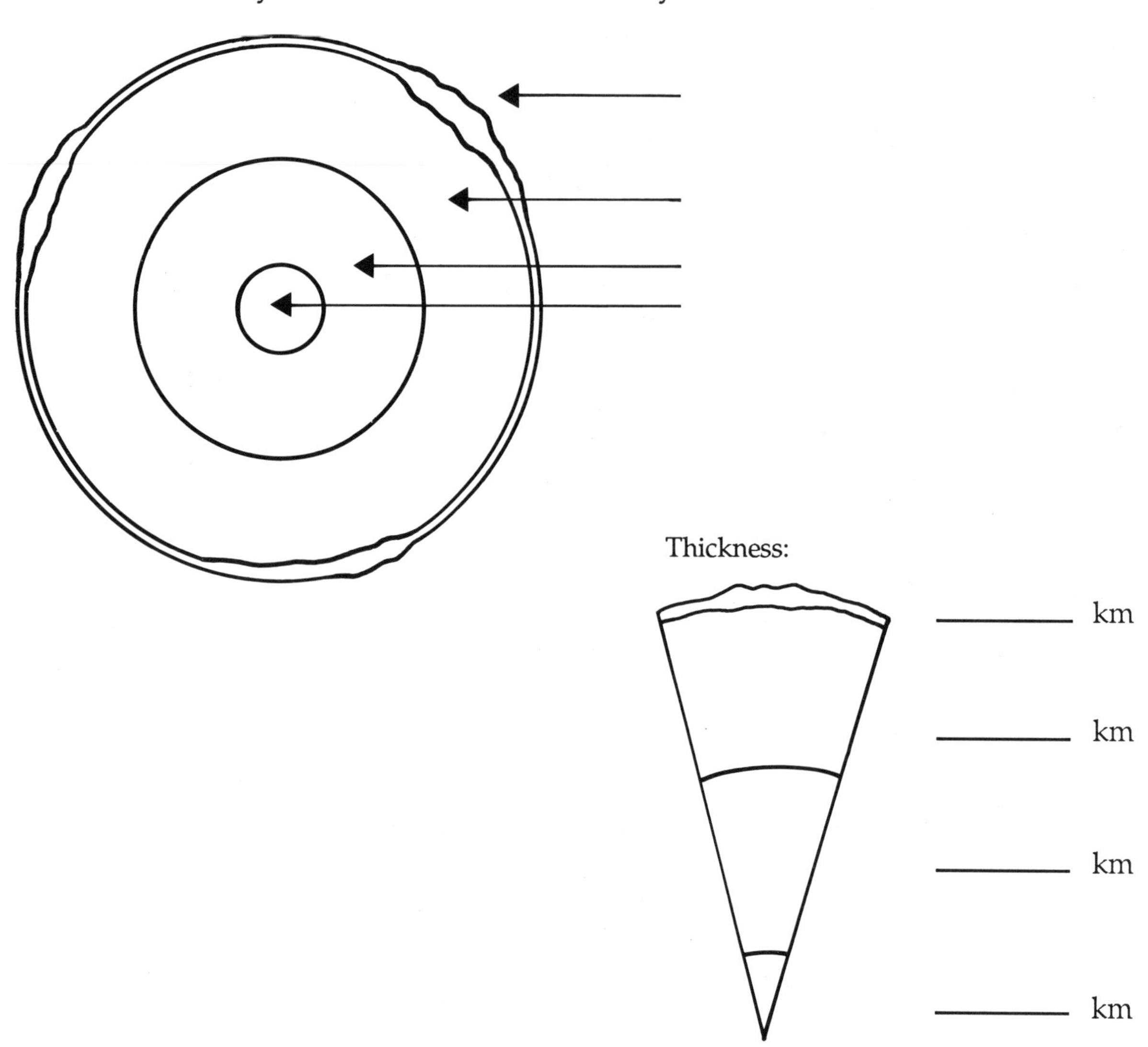

CHART THE EARTH'S INTERIOR

Layer	Thickness	Composition
______________	______________	______________________
______________	______________	______________________
______________	______________	______________________
______________	______________	______________________

Lesson 2

Do the Pieces Fit?

Recreating the supercontinent Pangea

Group Size
Individuals, with subsequent class discussion

Time Required
1 class period

Materials
For each student:
- copy of Handout 1, "Continent Jigsaw Puzzle," on page 19, and Handout 2, "Fossil Evidence Map," on page 20
- pencils and colored pens
- scissors
- glue
- 1 sheet colored construction paper

For the classroom (optional):
- large world map (preferably relief)
- colored paper and marking pens for making indicators of fossil evidence that go on the classroom map

Key Terms

continental drift	Pangea
Glossopteris	fossil
Lystrosaurus	glacier
glacial drift	mid-ocean ridge
glacial striations	

Instructional Goal
- To increase understanding of continental drift and how the continents may have fit together 250 million years ago.

Student Objectives
Students will:
- Recreate the supercontinent Pangea by fitting continental shapes together.
- Use fossil and glacial evidence provided by the teacher to piece Pangea together more accurately.
- Formulate hypotheses based on the fossil and glacial evidence.
- Describe in writing the evidence that supports the theory of continental drift.

Prerequisite Knowledge
Students should be familiar with maps of the world and should be able to identify and locate the continents.

Advance Preparation Time
About 15 minutes
- Review the "Background Information" and "Lesson Procedure."
- Gather materials.
- Put up a large world map if one is not already displayed.
- Duplicate the student handouts.
- Make overhead transparencies or copies of Figures 1 and 2, if desired, for class discussion.

Teacher Tips
- In addition to giving your students the "Fossil Evidence Map," which contains the relevant fossil and glacial evidence, you may wish to place this information on a large map posted in your classroom. Sketches of the fossils *Glossopteris* and *Lystrosaurus* have been included on the "Fossil Evidence Map" and can be used as markers for a large classroom map.
- It may be necessary to review basic map skills with your students, including the locations and names of the continents.

Background Information

The theory that the continents were once joined together and have slowly moved apart to their present positions has been proposed and forgotten many times during the last century. Little significance was given to the idea at first, primarily because there were few pieces of evidence that couldn't be explained by less extraordinary processes. In the early 1900s, Alfred Wegener and his follower Alex du Toit popularized the concept once again, but they lacked the crucial evidence to convince the rest of the scientific community.

Even with today's technology, actual measurement of plate and continental movement is difficult. However, a great deal of indirect evidence gives us clues about continental drift. Both plant and animal fossils provide strong supporting evidence for continental movement. Fossils of *Glossopteris*, an ancient plant, have been found in rocks of the same age in India, South America, Antarctica, Australia, and Africa. Fossils of *Lystrosaurus*, a plant-eating amphibious reptile about the size of a large dog, have been found in Antarctica, Africa, and India. Additional fossils of many other types of plants and animals have been found in widely separated landmasses, supporting the concept that these landmasses were once connected.

The rock record of 250 million years ago also includes the distinctive deposits known as glacial drift—rock and soil pushed by glaciers into identifiable deposits. Similar glacial drift deposits and associated directional glacial striations (grooves created by the movement of glaciers) have been found in South America, South Africa, Australia, Antarctica, and India, indicating that the continents have changed positions. The glacial deposits suggest that Africa and South America were further north and contiguous. When the continents are "moved together" in jigsaw-puzzle fashion it is much easier to see how a sheet of glacial ice may have covered part of Pangea and left the associated deposits.

In addition to this evidence, the shape of the continents (particularly South America and Africa) and their fit with the spreading mid-Atlantic ridge provide support that the supercontinent Pangea did indeed exist about 250 million years ago. (See the "Fossil Evidence Map," which shows the location of these pieces of fossil and glacial evidence.)

Lesson Procedure

Step 1:

As an introduction to the lesson, have your students look at a world map and discuss any spatial relationships they notice. Encourage them to look at the continent shapes.

Step 2:

Give each student a copy of Handout 1, construction paper, scissors, and glue. Have students cut out the continents and arrange them on construction paper as they are on a present-day world map.

Ask students to see if they can fit the continents together like pieces of a jigsaw puzzle. Encourage students to compare their maps. Discuss their findings and the maps they made.

Discuss what facts or information might help them decide which way to put the continents together. Talk about the fossils and glacial evidence that have been found, as described in "Background Information." Go over each of the "Key Terms."

Step 3:

Pass out the "Fossil Evidence Map" to each student, which shows where evidence has been found. As described in "Teacher Tips," you can also post this information on a large world map.

Have your students place their cut-out continents back as they are on a present-day world map. Students should transfer the facts from the "Fossil Evidence Map" to their cut-out continents.

Using the fossil and glacial evidence, your students should attempt to fit their cut-out continent pieces together again (but do not glue them together yet). The evidence should enable them to produce a better fit similar to Figure 2 on the next page.

Have your students compare their maps to a map of Pangea (see Figure 1 below). Ask your students if they obtained configurations that are similar to the geologists' reconstruction of Pangea. You might use transparencies of these figures on an overhead projector, or make copies for handouts. Discuss how the fossil and glacial evidence improved or verified their first attempt at piecing the continents together based only on their shapes. Ask students if they can think of any other way that animals, plants, and glaciers could have left this fossil record if the continents were not joined together.

Students should now glue their maps of Pangea to construction paper. The maps can then be displayed.

Step 4:
In addition to making a reasonable map for Pangea based on the given evidence, have students write a paragraph describing how the evidence supports their maps of Pangea.

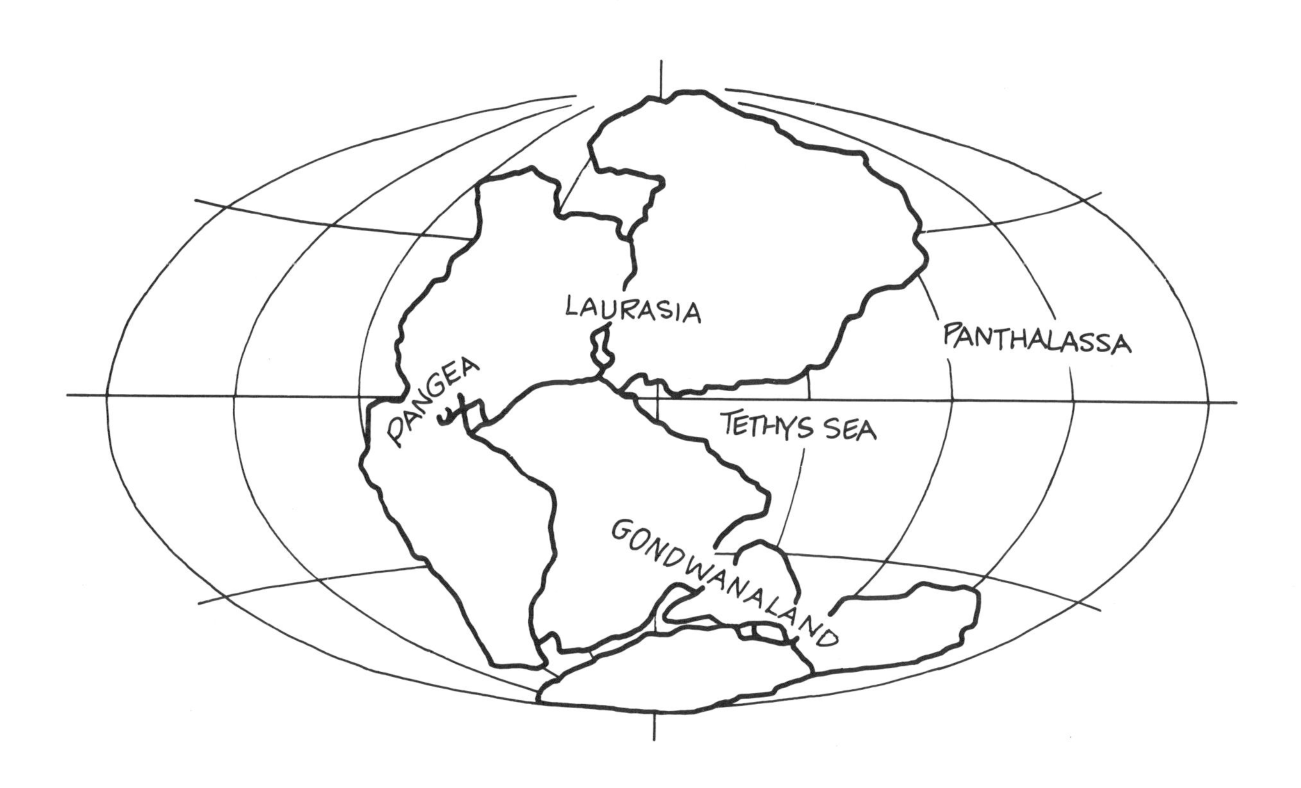

Figure 1. Geologists' reconstruction of the supercontinent Pangea.

Enrichment Activities

Students can:

- ☐ Find out what other fossil evidence of continental drift and Pangea has been found.
- ☐ Research Alfred Wegener and Alex du Toit.
- ☐ Find out who proposed the theory of continental drift before Wegener. Who was the first?
- ☐ Make a three-dimensional model of Pangea, including symbols for the fossil and glacial evidence.

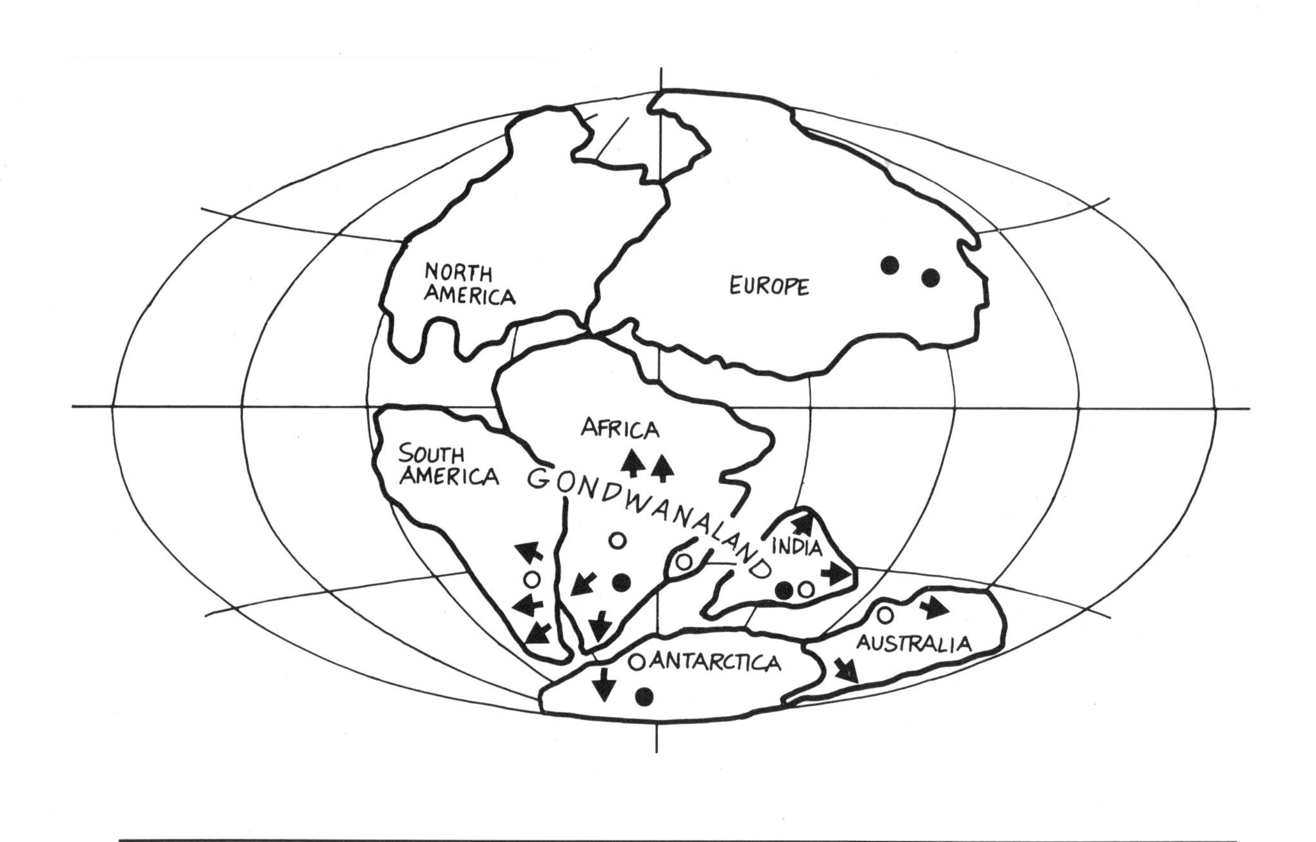

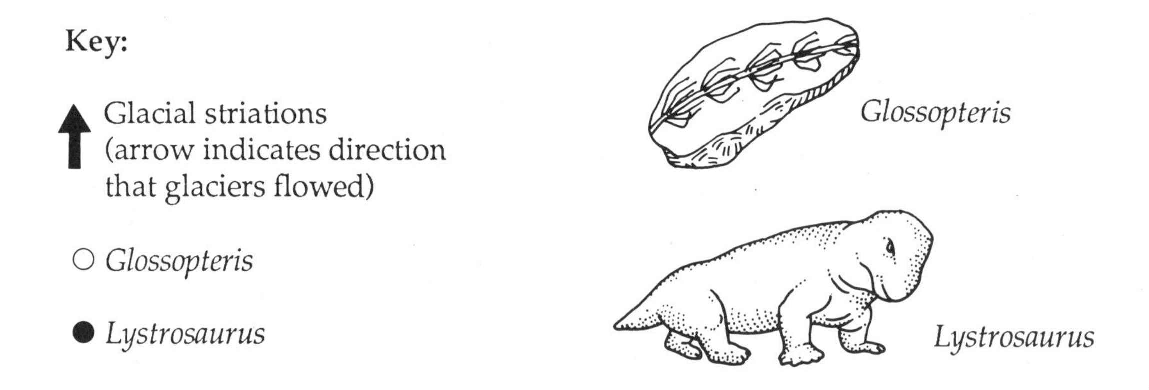

Figure 2. Fossil and glacial evidence produce this configuration of Pangea.

Continent Jigsaw Puzzle

1. Cut out and label the continent shapes.
2. Arrange them as they appear on a world map today.
3. Try to make the continents fit together, like pieces of a jigsaw puzzle.
4. Draw the fossil and glacial evidence from the "Fossil Evidence Map" on your continent shapes. Try to improve the fit of your puzzle using this additional information.
5. Glue the pieces to a sheet of construction paper.

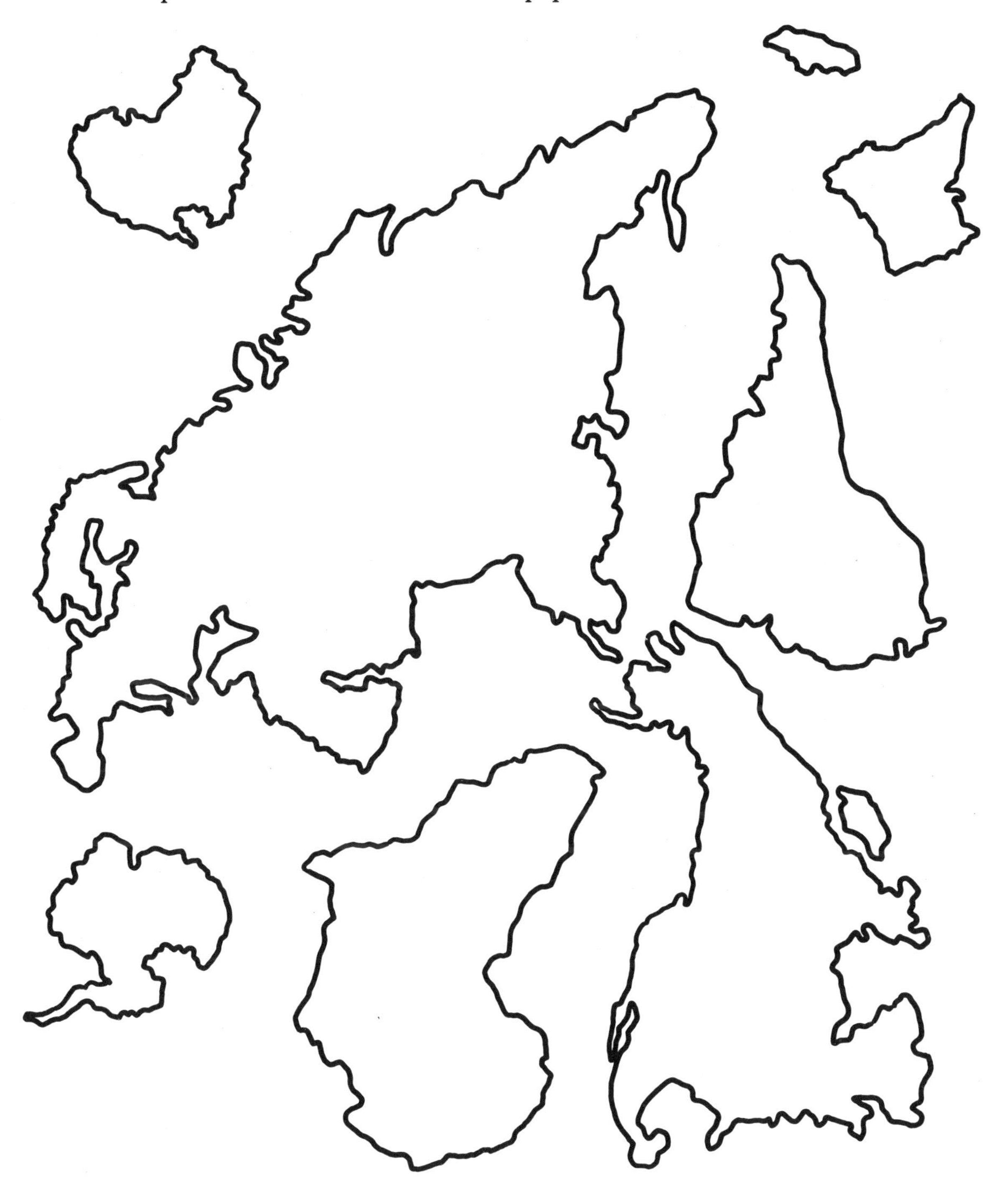

Fossil Evidence Map

Fossils of plants and animals that lived about 250 million years ago have been found on continents that are now separated by vast oceans. Evidence of ancient glaciers from the same time period has also been found. Transfer the information on this map to your continent shapes.

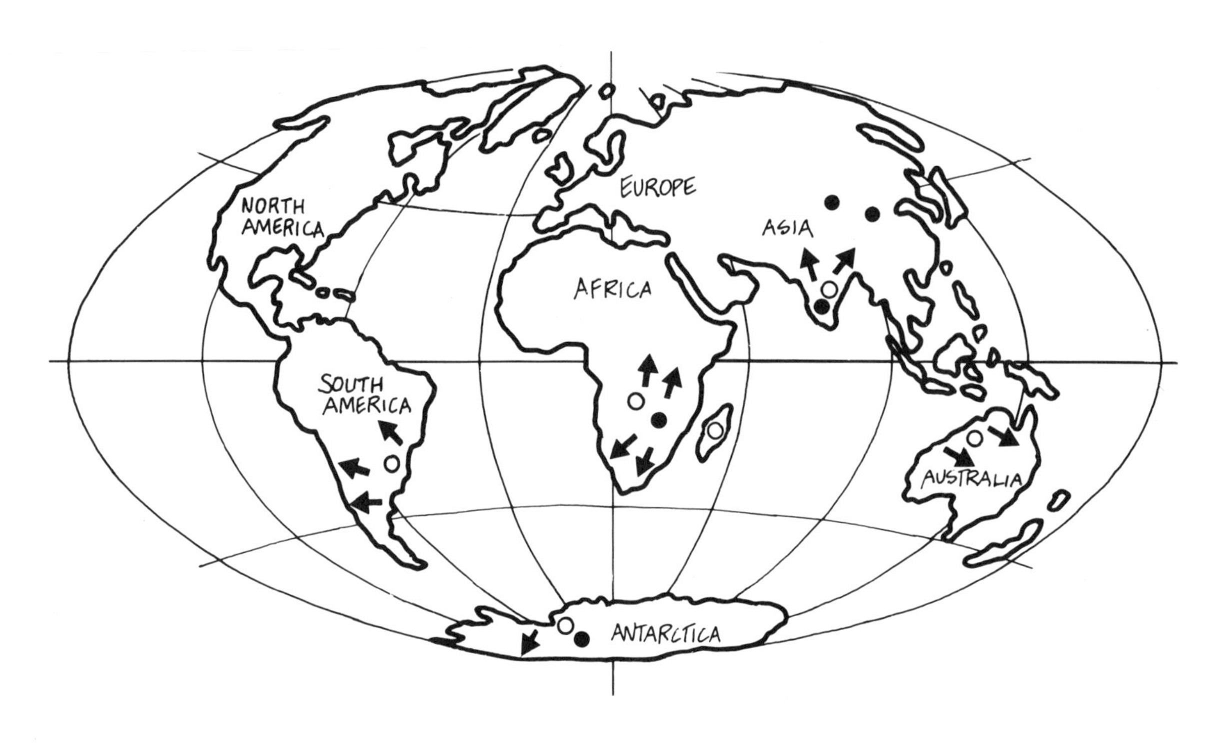

Key:

 Glacial striations (arrow indicates direction that glaciers flowed)

○ *Glossopteris*

● *Lystrosaurus*

Glossopteris

Lystrosaurus

Where Have They Been?

Tracing continental movement during the last 250 million years

Group Size
Individual students

Time Required
1 class period

Materials
- copy of Handout 1, "Continental Drift Sequence," on page 25, and Handout 2, "Where Are We Going?" on page 26, for each student
- colored pencils and pens
- white paper for tracing (optional)

Key Terms

Pangea	Laurasia
Panthalassa	Tethys Sea
Gondwanaland	

Instructional Goal
- To develop an understanding of how the continents have moved in the past to create the earth as it is today.

Student Objectives
Students will:
- Observe on a sequence of maps the directions of continental movement during the last 250 million years.
- Describe in writing how the continents have moved, based on these maps.
- Predict the configuration of the continents 50 million years from now, based on data about the directions the continents are moving.
- Diagram their predictions.

Prerequisite Knowledge
Students should have basic knowledge of the theory of plate tectonics and continental drift, as introduced in the previous lessons.

Advance Preparation Time
About 15 minutes
- Review the "Background Information" and "Lesson Procedure."
- Gather materials.
- Duplicate the student handouts.
- Make a transparency of the "Future Possibilities" sheet at the end of this lesson, or prepare copies for students.

Teacher Tips
- Although your students will focus on the movement of the *continents* in this activity, it is important for them to understand that the continents rest on *plates*. Since it is the plates that move, students may find it somewhat difficult to predict the future configuration of the continents. The chart on Handout 2 showing the direction of movement for the major plates is given to help students with their predictions.
- If possible, obtain a copy of *National Geographic*, August 1985, and *Discover*, November 1982. Articles in these issues show several computer-generated possibilities for future continental arrangement.
- To enhance class discussions, have several types of maps and globes available (different projections, relief, physical, etc.).

Background Information

The earth's great landmasses have been traveling over the globe for hundreds of millions of years. (Some geologists extend this back as far as 4 billion years.) By analyzing the magnetic patterns recorded in the lava spreading out from the mid-ocean ridges, and by matching fossils and rocks from different continents, scientists have been able to trace the patterns of continental movement for the last 600 million years. This lesson focuses on the earth as it was about 250 million years ago (during the Permian Period), when all the continents were connected in a supercontinent known as Pangea (meaning "all land") and the remainder of the globe was covered by the great ocean Panthalassa. The Tethys Sea existed along the eastern shores of Pangea; the Mediterranean Sea is what remains of the Tethys Sea today.

According to plate tectonic theory, Pangea first broke up into two large landmasses—Gondwanaland (containing the present-day landmasses of India, Australia, Africa, and South America) and Laurasia (containing North America, Europe, and Asia). At the same time, the mid-Atlantic ridge began opening between North America and Africa, forming the Atlantic Ocean. About 165 million years ago, the northern Atlantic and Indian Oceans took shape and the southern Atlantic Ocean widened. Australia was still attached to Antarctica at this time. About 100 million years ago, several major changes occurred that created today's continental configuration—Australia was torn away from Antarctica, North America and Eurasia separated, and India ran into Eurasia, creating the Himalayas. Although the plates on which the continents rest move only a few centimeters each year, over millions of years the continents have traveled thousands of miles. The continents continue to move today, and based on the known direction and speed of each plate's travel, scientists have created computer-generated maps of the future locations of the continents. Thus, geologists are able to look forward into the future as well as back into the past.

During the next 100 million years, according to one possible model, the Pacific Plate will carry a piece of California north to Alaska. The movement of the African Plate will create a mountain range in the Mediterranean, leaving a large piece of the East African coast behind. Australia will push into Indonesia, creating an extension of the Himalayas. A deep ocean trench will form along the east coast of North and South America. About 150 million years from now, this trench will begin to shrink the size of the Atlantic Ocean. The Eurasian-African landmass will slowly tilt clockwise, exposing western Europe to the arctic cold. In 250 million years, a new Pangea may exist.

This sequence of events is only one possibility. It is also exciting to explore the idea that plant and animal species now separated by oceans may be able to interact someday.

Lesson Procedure

Step 1:

As an introduction, discuss with your class the idea that the plates have been gliding and colliding over the face of the globe for many millions of years. Scientists have developed a possible sequence of past continental movement. Emphasize that the plates are still moving—earthquakes and volcanoes are just two indicators that the earth's crust is volatile. This should lead to a question or comments from your students about the future configuration of the continents.

Step 2:
Give copies of the two handouts to each student. Students should examine the patterns of continental drift in the series of maps on Handout 1, using colored pens or pencils to distinguish the various continents. Be sure to point out the various names that geologists have given to the ancient landmasses and oceans (see "Key Terms"). Students then record their observations on Handout 2. Ask students to discuss their observations. Are the continents all moving in the same direction? Different directions? If so, which continents have moved away from each other, and which continents have moved toward each other?

Step 3:
Ask your students to predict the position of the continents 50 million years from now, using the information given on the Handout 2. Students should draw their prediction maps in the space provided. As an optional aid to constructing their prediction maps, students may trace the present-day continent shapes from Handout 1.

After each student has drawn a prediction map, ask your class to compare their predictions with the diagrams presented on the "Future Possibilities" sheet. Show a transparency of this sheet on an overhead projector, or hand out copies. What differences and similarities do your students observe between their maps and those on the transparency? What effects will these new continental relationships have on plant and animal species?

You may also wish to have your students compare their maps with the predictions made in *National Geographic* and *Discover* (see "Teacher Tips" for issue dates). Discuss reasons for the predictions made by the scientists in these additonal sources and by your students.

Step 4:
Ask your students to write a paragraph or two describing how they arrived at their prediction maps. As a check for understanding, you may also wish to have your students write a description of continental movement during the last 250 million years.

Enrichment Activities

- ☐ Students can do further research on continental drift in the past (see *National Geographic*, "Earth's Dynamic Crust," August 1985) or possible continent configurations in the future (see *Discover*, "The Shape of Tomorrow," November 1982).
- ☐ Using Handout 1, rearrange the sequence of the maps so they are out of order, make copies for your students, and have them number the maps in the correct sequence.
- ☐ Obtain a 5-inch styrofoam ball and paint it blue to represent a globe. Cut continent shapes out of green felt using the pattern shown on page 24. The continents can then be placed on the globe to represent different configurations. This model enables students to see the patterns of gradual continental movement more clearly.

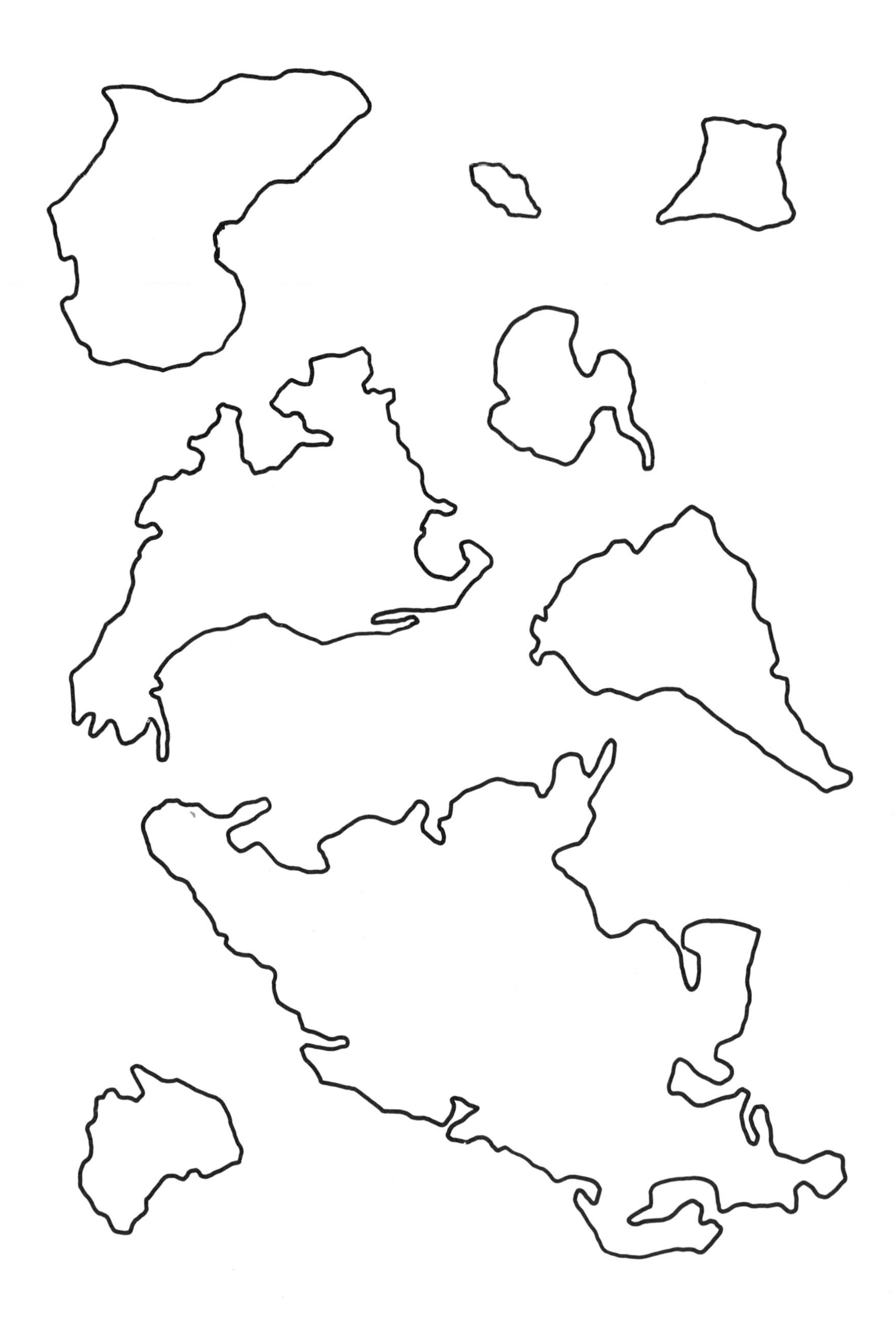

Pattern pieces for felt continent shapes (see Enrichment Activities).

Continental Drift Sequence

Observe how the continents have moved during the last 250 million years. Have they moved in the same direction or in different directions? Color each continent with a different color, keeping the same color for all the maps, so it is easier to see how the continents have moved.

250 million years ago

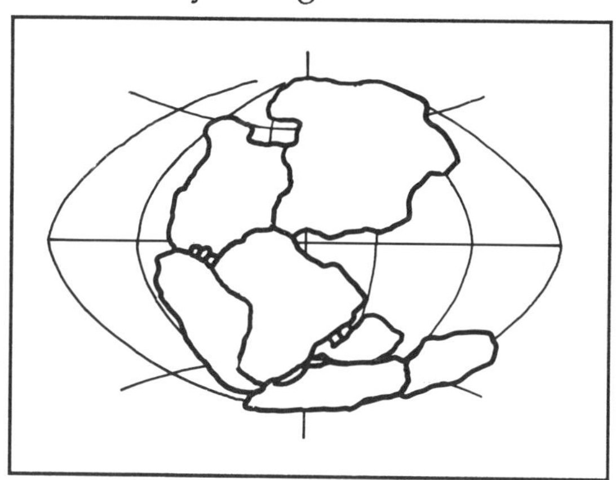

180 million years ago

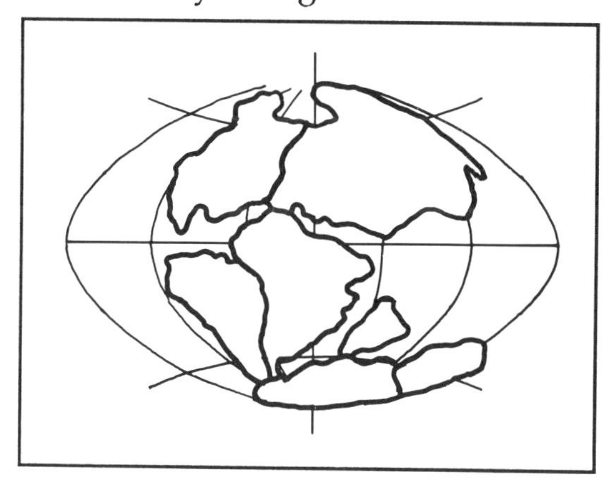

135 million years ago

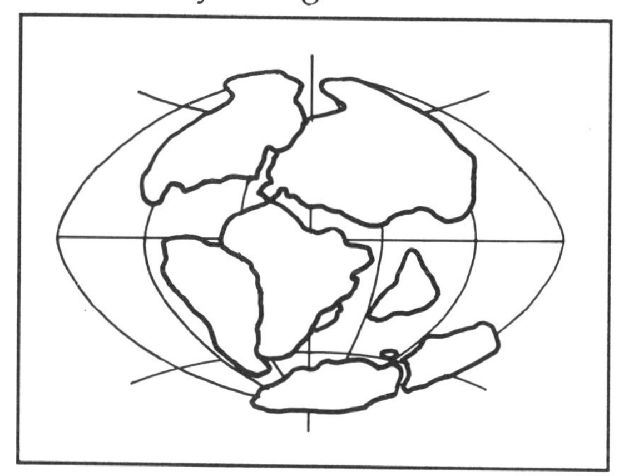

65 million years ago

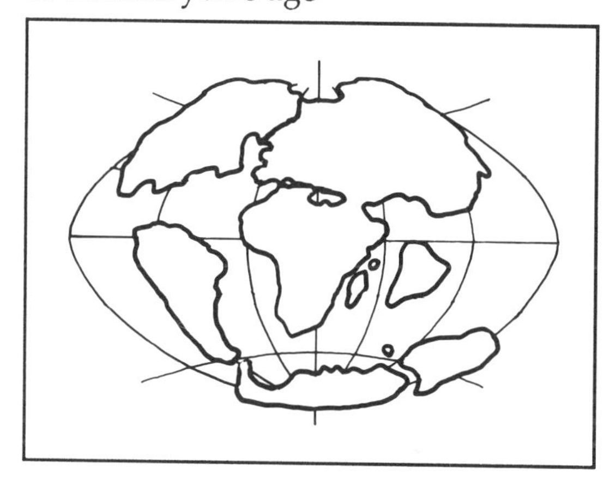

Today

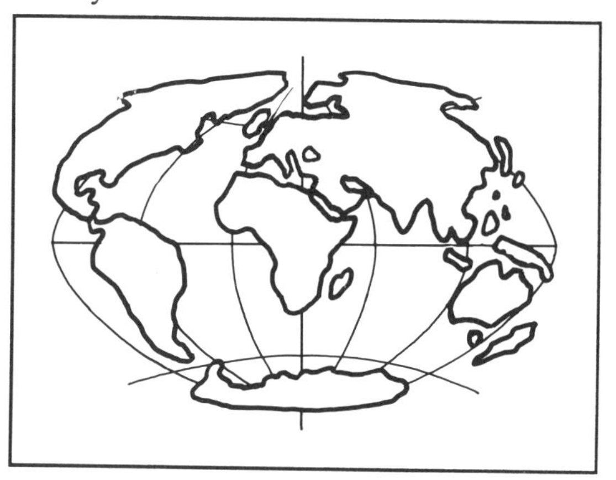

50 million years in the future

Where Are We Going?

Using the maps on Handout 1, write your observations about how the continents have moved during the last 250 million years. What directions have they moved? Which continents separated? Which continents came together?

Observations

250-180 million years ago:

180-135 million years ago:

135-65 million years ago:

65 million years ago-today:

Predict

How will the earth look 50 million years in the future? Using the table that lists the direction of continental movement, draw your prediction on the map below.

Continent	Direction
North America	northwest
South America	west
Eurasia	east
Africa	east-northeast

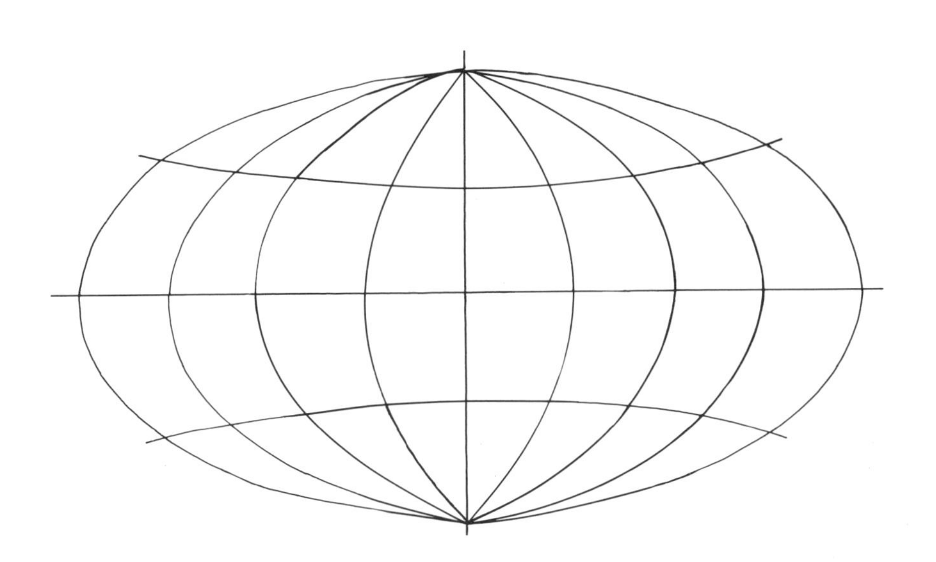

Future Possibilities

Scientists don't all agree on how the earth will look in the future, but several possibilities are shown below. What would you have to know in order to predict how the earth will look millions of years from now? What would happen if some of these things (variables) changed?

50 million years from now

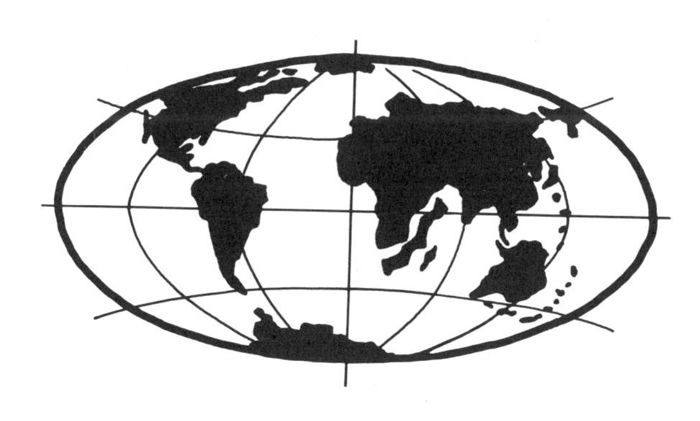

50 million years from now

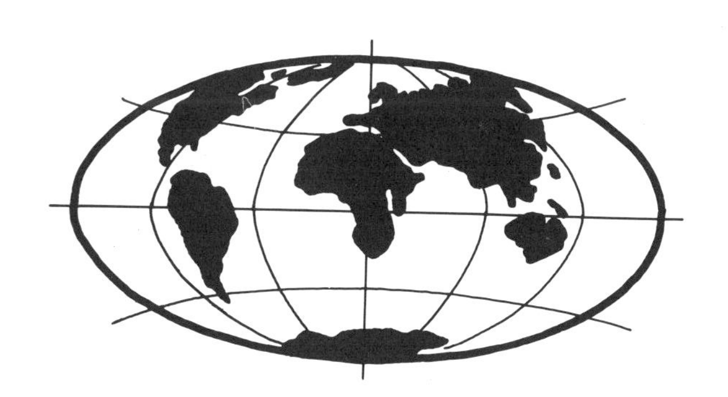

100 million years from now

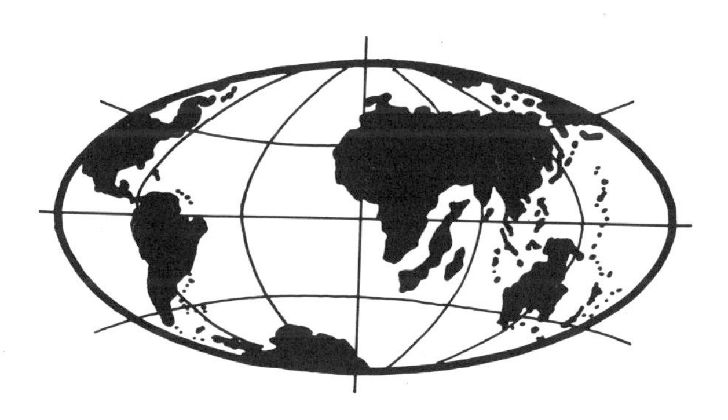

150 million years from now

Lesson 4

Where Are the Plates?

Locating plate boundaries through earthquake and volcano zones

Group Size
3-5 students, or individuals

Time Required
2-3 class periods

Materials

- one large world map for each group, or a copy of the "Individual World Map" on page 33 for each student
- copy of Handout 1, "Find It!" on page 34; Handout 2, "The Earth's Plates," on page 35; and Handout 3, "Plate Boundaries Fact Sheet," on page 36, for each student
- transparency of the "Answer Key Map" on page 32 (optional; or use as handouts)
- miscellaneous supplies for making map symbols—construction paper, colored pens and pencils, glitter, yarn, etc. (students can bring from home)
- a collection of earth science textbooks, encyclopedias, magazines, etc., with information on plate tectonics, earthquakes, and volcanoes

Key Terms

seafloor spreading	trench
plate tectonics	plates
transform fault	mid-ocean ridge
diverging boundary	earthquake
converging boundary	volcano

Instructional Goals

- To understand why earthquakes, volcanoes, and landforms such as trenches and mid-ocean ridges occur most often along plate boundaries.
- To recognize the three types of plate boundaries and their associated geologic features.

Student Objectives

Students will:

- Formulate hypotheses based on evidence found in earth science books.
- Indicate on a world map the location of earthquake zones, volcanic activity, trenches, mid-ocean ridges, and various mountain ranges.
- Discuss why volcanoes and earthquakes occur where they do.
- Describe each type of plate boundary.

Prerequisite Knowledge

Rudimentary information about volcanoes and earthquakes; basic map skills.

Advance Preparation Time

30-45 minutes

- Review the "Background Information" and "Lesson Procedure."
- Gather materials, especially a variety of earth science resource books.
- Ask your students to bring from home items for making map symbols (see "Materials"), thus making their maps more personalized.
- Duplicate the student handouts and make transparencies.

Teacher Tips

- This activity may be done either by individual students or with groups of 3-5 students. If you decide to use groups, choose group members to maximize cooperation. During the activity, observe student interaction. I usually include group cooperation as part of my grading for this activity.
- Post several different types of maps and globes. Ask students to compare physical maps, relief maps, and various projections.

Background Information

According to the theory of plate tectonics, the earth's crust is broken up into big blocks called plates. A plate consists of a thin crustal layer that rests on top of a thick layer of dense mantle material. The plates are collectively referred to as the lithosphere (the Greek *lithos* means "stone"). The earth's surface consists of approximately 20 plates, some of which contain both the thick continental crust and the thin oceanic crust, and some of which contain only thin oceanic crust. The major plates are the African, South American, North American, Pacific, Indo-Australian, and Antarctic plates. Some of the smaller plates are the Cocos, Nazca, Philippine, and Caribbean plates. Handout 2 shows a map of the locations of these plates.

The plates move toward or away from each other as they float on the partially molten layer of the upper mantle known as the asthenosphere (the Greek *asthenes* means "weak"). Earthquakes and volcanoes are caused by these shifting plates, and tend to be located primarily near plate boundaries. Scientists made the first approximations of the plate boundaries by plotting earthquake and volcanic activity, just as your students will do in this lesson. Extensive oceanic and earthquake research has enabled scientists to compile a map of the earth's plate boundaries (see Handout 2).

Dramatic events can happen at the boundary between two plates. The existence of these boundaries gives us an explanation for the earthquake and volcanic activity that occurs somewhere in the world every day. Depending on the type of boundary, the rock formations in the area, and the speed at which the plates are moving, various types of landforms are created.

There are three types of plate boundaries—diverging, converging, and transform (see the "Plate Boundaries Fact Sheet" on page 36 for diagrams). A *diverging* boundary occurs between two plates that are separating, or moving away from each other. Molten material rises to the surface along a ridge on the ocean floor, producing new hardened rock. This new rock is pushed away from the center of the mid-ocean ridge as new molten lava continuously rises to the surface. The hardened rock forms new crustal material in this process, which is known as seafloor spreading. When examining a physical or relief map of the world's oceans, it is easy to see the mid-ocean ridges that extend around the globe like the seams of a baseball.

Converging plate boundaries occur where two plates are coming together. One form of converging boundary produces a subduction zone in which an ocean plate buckles underneath a thicker continental plate. This subduction produces a deep trench where the oceanic crust is pushed downward toward the mantle. The oceanic crust melts as it is pushed deeper into the mantle, thus "recycling" the earth's crust. Volcanic activity, earthquakes, and deep trenches are usually found along this type of plate boundary. The second form of converging boundary is found when two continental plates come together. This type of boundary usually produces very high mountain ranges, such as the Himalayas.

The third type of plate boundary is a *transform fault*. These faults are found when two plates slide past each other. A well-known example of this type of boundary is the San Andreas Fault along the west coast of North America. For diagrams of each of these types of plate boundaries, see the "Plate Boundaries Fact Sheet" at the end of this lesson.

The existence of plates and their relative movement are well-documented parts of plate tectonic theory. Continental drift is just another form of movement in our ever-changing universe.

Lesson Procedure

Step 1:
To introduce the lesson, discuss the most recent volcanic eruption or earthquake in the news. Have one of your students find its location on a large world map. Locate a couple of other famous earthquakes or volcanoes, or those your students suggest.

Step 2:
This activity may be done by individual students or with groups of 3-5 students. If you are having students complete the activity individually, distribute a copy of the "Individual World Map" on page 33 to each student. Alternatively, divide your class into groups, and give each group a large world map.

Next, distribute copies of Handout 1 to each student. Handout 1 asks students to locate a number of geologic features, earthquakes, and volcanoes, and to show them symbolically on a world map.

Using the earth science resource books you have collected, students should locate on their world maps the landforms and locations listed on Handout 1. You may want to add other locations to the handout that you feel are relevant, such as the school's location or a geologically active area in the news.

Step 3:
Students should decide how to indicate the various types of landforms or geologic activities on their maps; e.g., what symbol to use to denote an earthquake, or what materials to use to make a symbol for a mountain range. For example, a small cone made of brown or orange construction paper could symbolize a volcano. Each map should also have a legend or key to describe the symbols.

Students in groups will find that this activity is easier if each group decides on a leader who delegates individual tasks for making the symbols.

I evaluate my students' maps based on neatness, completion, correctness, and, if applicable, group cooperation. When using groups, all students in a group receive the same grade for their map based on these criteria.

In the next step, students add the plate boundaries to their individual maps.

Step 4:
When all the maps have been completed, discuss the locations and relationships among the geologic features and activities. Ask your students if they noticed any patterns. Are there any concentrations of symbols in certain areas? What do these geologic activities indicate? Why would these geologic events happen most often in these areas?

Discuss the concept of plates and the three different types of plate boundaries as presented in the "Background Information," as well as each of the "Key Terms." Be sure to explain the relationship between plates and continents. Display an overhead transparency of the "Plate Boundary Fact Sheet," or hand out copies of this sheet instead. Have your students look on their maps for sections of crust they think might be plates. Students should understand what types of geologic features or activities are clues to the associated type of plate boundary.

Give each student a copy of Handout 2, which shows the plate boundaries as we know them today. Have your students add the plate boundaries to their maps. Students should also add the names of the major plates and the direction of plate movement (from Lesson 3).

The student maps can be made self-correcting by making a transparency of the "Answer Key Map" on page 32. This map shows the locations of the geologic features listed in Handout 1. Display the transparency on an overhead projector, or have students pass the transparency around and individually place their maps underneath it as a self-check.

Step 5:
After the maps have been corrected, ask your students to find examples of each of the three types of plate boundaries on their maps. What occurs at the mid-ocean ridges and at the trenches? What is seafloor spreading? Where is the youngest and the oldest crust? How are transform faults produced?

Step 6:
For oral discussion, I have several individual students, or one member from each group, describe their maps and how they determined the plate boundaries. I also have my students compose a paragraph explanation of why volcanoes and earthquakes occur where they do.

Enrichment Activities
Students can:
- ☐ Research a particular plate.
- ☐ Make a three-dimensional model of a plate boundary cross-section.
- ☐ Research the latest scientific information about the direction and speed of each plate's movement.
- ☐ Find out about a famous plate boundary, such as the San Andreas Fault.
- ☐ Investigate the closest plate boundary to your school.
- ☐ Find out how the Hawaiian Islands are related to plate tectonics. Why are these volcanoes located so far away from a plate boundary?
- ☐ Research Iceland's relationship to the dynamics of plate tectonics.
- ☐ Investigate projections of what will happen to California during the next 150 million years, then make a model to show one prediction.

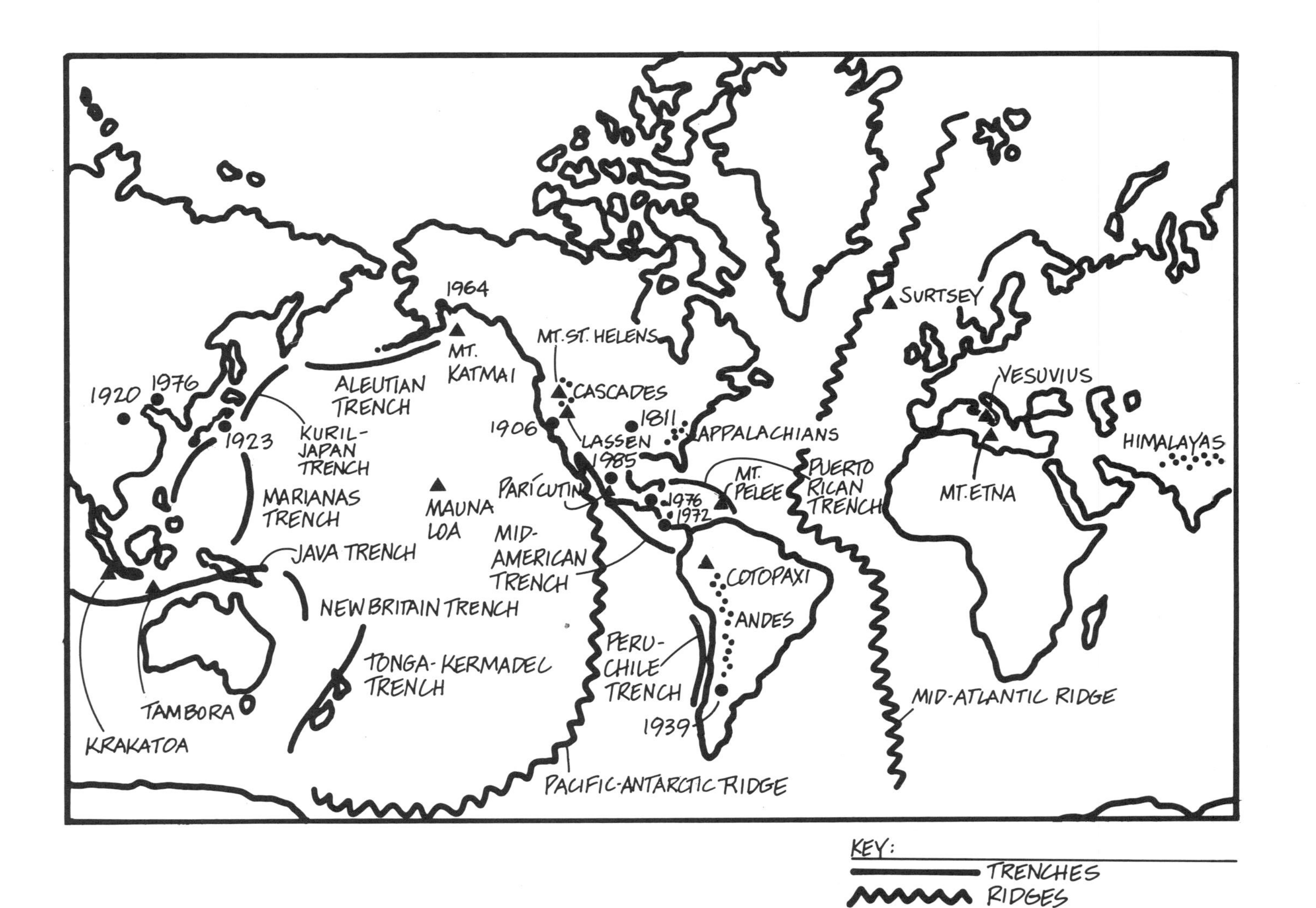

Answer Key Map for geologic features listed on Handout 1.

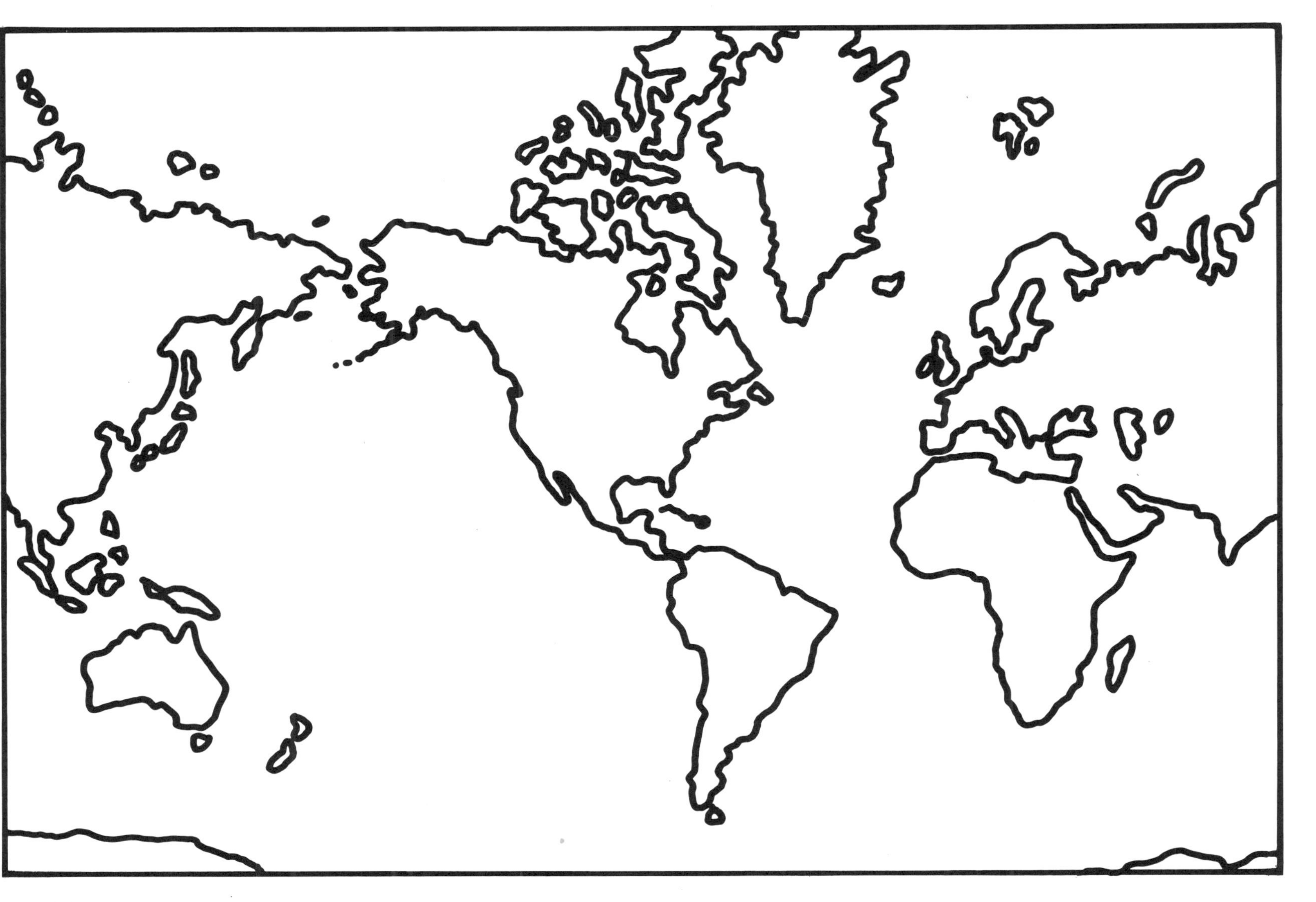

World Map

Find It!

Indicate the locations of the following geologic features and activities on a world map. Decide what symbol to use to indicate each item, and what materials to use to make your map symbols. Be sure to put a legend on your map.

Trenches

1. Aleutian
2. Tonga-Kermadec
3. Peru-Chile
4. Kuril-Japan
5. Puerto Rico
6. Java
7. Marianas
8. Mid-American
9. New Britain

Mid-ocean Ridges

1. Pacific-Antarctica Ridge
2. Mid-Atlantic Ridge

Mountain Ranges

1. Himalayas
2. Andes
3. Cascades
4. Appalachians

Major Earthquakes

1. 1811 New Madrid, Missouri (strongest recorded in United States)
2. 1906 San Francisco, California (500 people died)
3. 1920 Gansu Province, China (100,000 died)
4. 1923 Toyko, Japan (142,000 died)
5. 1939 Chillan, Chile (30,000 died)
6. 1964 Anchorage, Alaska (131 died)
7. 1972 Managua, Nicaragua (10,000 died)
8. 1976 Guatemala (22,000 died)
9. 1976 Tangshan, China (800,000 died)
10. 1985 Mexico City, Mexico (10,000+ died)

Volcanoes

1. Cotopaxi, Ecuador (19,347 feet high, last eruption in 1975)
2. Mont Pelee, Martinique (1902 eruption killed 38,000 people)
3. Krakatoa, Indonesia (1883 eruption created tsunamis, drowned 36,000)
4. Lassen Peak, California (last eruption was in 1921)
5. Mauna Loa, Hawaii (world's tallest, rises 30,000 feet above ocean floor)
6. Mt. Etna, Sicily (an active volcano over 11,000 feet high)
7. Mt. Katmai, Alaska (1912 eruption created the Valley of 10,000 Smokes)
8. Tambora, Indonesia (1815 eruption had energy of 6,000,000 atom bombs)
9. Parícutin, Mexico (rose out of a farmer's field in 1943)
10. Surtsey, near Iceland (began forming an island in 1963)
11. Vesuvius, Italy (buried Pompeii in A.D. 79)
12. Mount St. Helens, Washington (violent eruption on May 18, 1980)

The Earth's Plates

Shown below are the boundaries of the earth's plates. Notice how the plates vary in size. Find an example of the three different types of plate boundaries and label them.

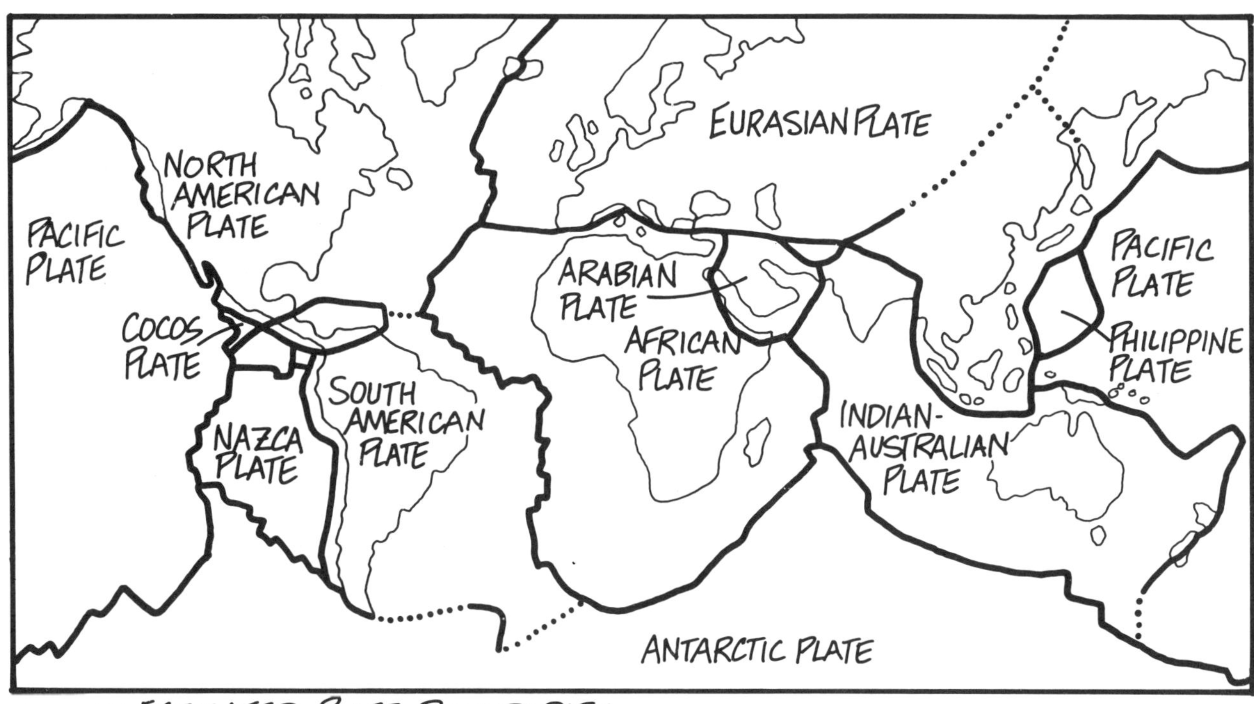

Plate Boundaries Fact Sheet

Plates meet at boundaries. The interaction of moving crustal plates produces a variety of geologic activities, such as earthquakes and volcanoes, and geologic features, such as mid-ocean ridges and trenches.

There are three types of plate boundaries:

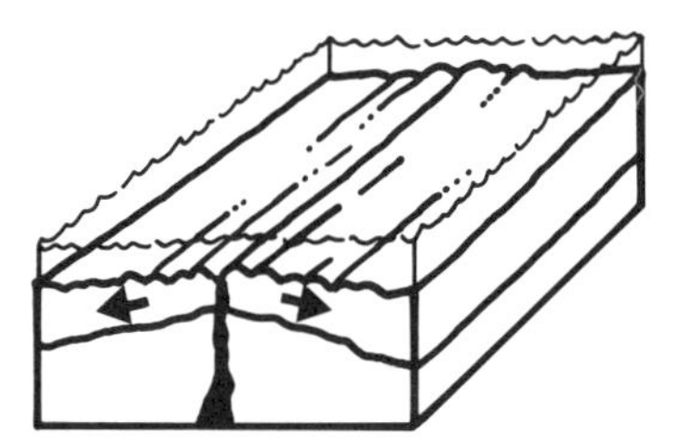

Divergent

- Occurs between plates that are *separating*
- Indicated by *seafloor spreading*
  - Caused by molten rock rising to the surface and hardening
  - Hardened rock is pushed outward by new molten lava
  - Moves about 5 cm per year
- Examples: East-Pacific Rise, Mid-Atlantic Ridge

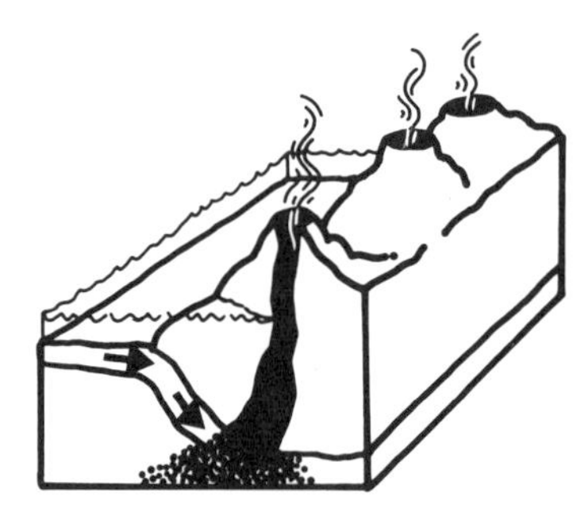

Convergent

- Occurs where plates *come together*
- *Subduction*
  - Occurs where oceanic and continental plates come together
  - Thinner oceanic crust buckles downward under continent, is pushed into mantle, and melts
  - Usually forms volanoes inland and parallel to the trench
  - Example: Aleutian Trench
- *Collision*
  - Occurs where two continental plates come together
  - Usually forms high mountain ranges
  - Examples: Himalayas, Appalachians

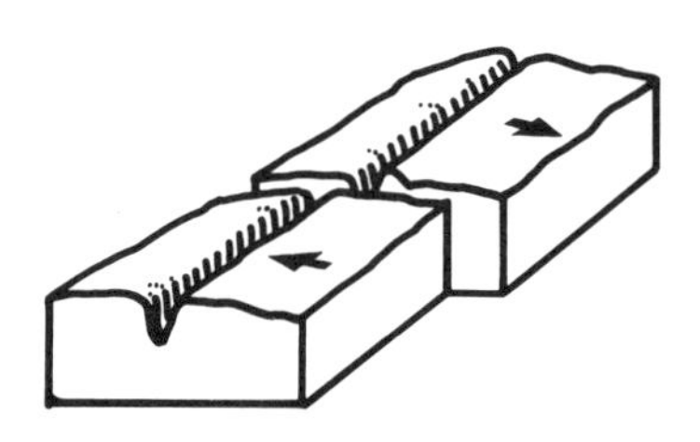

Transform Fault

- Occurs where plates *slide past each other*
- Results in localized deformation of landscape, low mountains, valleys
- Example: San Andreas Fault at boundary of Pacific Plate and North American Plate

Lesson 5

Ride the Convection Coaster

A laboratory investigation of movement in heated liquids

Group Size
2-4 students

Time Required
1 class period

Materials
For each student:
- copies of Handout 1, "Ride the Convection Coaster," on page 40, and Handout 2, "Use What You've Learned," on page 41
- colored pencils

For each group:
- 1 liter of water
- food coloring (divide a 1-oz. bottle between 3 groups; have several colors available)
- eyedropper
- clear glass loaf pan (Note: Although glass baking dishes can also be used, glass loaf pans work better for this experiment because of their depth.)
- 2-3 beakers or glasses to use as stands for the loaf pan
- candle (votive candles work well) and matches
- 1 beaker filled with water (for safety when working with matches)

Key Terms
convection
convection cell

Instructional Goals
- To promote understanding of the process of heat flow inside the earth.
- To relate this process of heat flow, or convection, to plate tectonics.

Student Objectives
Students will:
- Diagram a convection cell based on their observations of the movement of food coloring in heated liquids.
- Observe how changing the location of the heat source affects the flow of the colored liquid, then diagram the liquid's movement.
- Write about the relationship between plate tectonics and the movement of the heated liquids in the experiment.

Prerequisite Knowledge
Students should be aware of the concepts of plate tectonics as introduced in the previous lessons. In addition, students should be familiar with the laboratory skills of lighting a match safely and using an eyedropper accurately.

Advance Preparation Time
30 minutes
- Review the "Background Information" and "Lesson Procedure."
- Gather materials.
- Duplicate the student handouts.
- Put the food coloring into cups or beakers to make it easy for students to use their eyedroppers.
- Ask students to bring clear glass loaf pans from home in order to reduce materials cost.

Teacher Tips
- To reduce the class time required for set up on the day of the lab, have the materials already sorted and distributed to the student groups.
- Use several different colors of food coloring, preferably darker colors such as red, blue, and green.

Background Information

The basic concepts of plate tectonics received general scientific acceptance in the 1960s. After verifying the existence of the crustal plates and their associated movement, scientists focused on discovering the mechanism that moves these huge blocks of rock. Scientists believe that the driving force for the movement of the crustal plates is the process of convection within the mantle. Convection is caused by unequal heat distribution within this plastic-like layer.

A simple analogy that explains convection is the movement of boiling water in a pot. The warmest water rises to the surface, where it cools and then sinks back to the bottom, only to be reheated again. This is the circular motion of a convection *cell*, that is, an area where convection takes place (see Figure 1 on page 39). It is thought that the heated rock in the mantle moves in the same way, with hot material rising to the surface at the mid-ocean ridges. The melted rock hardens into new crust once it reaches the surface. It is then pushed outward as new molten material continues to rise at the mid-ocean ridges. Eventually the ocean lithosphere is pushed back into the mantle when it reaches a subduction zone. Scientists are currently using information recorded by digital seismometers to determine the location and size of the mantle's convection cells. The results of their research will produce a much clearer picture of what's going on beneath our feet.

Lesson Procedure

Step 1:
Distribute copies of the handouts and the materials, except the water, to each group. After the groups have set up their pans as shown on Handout 1, one student or the teacher should pour the water into each pan. *Very* cold water improves the experiment's results.

After the water has settled, students should follow the instructions on Handout 1. Remind your students to look through the *side* of the pan to see the movement of the food coloring.

Step 2:
Encourage students to try as many variations of the experiment as they can, such as putting the food coloring close to the heat source, far away from the heat source, using two heat sources, using two different dye colors, etc. This will enable students to get a more complete picture of how convection works.

Students should diagram their variations on Handout 1 and describe their results. Ask students to predict how the food coloring will respond when the heat source is placed in different positions. How do their predictions compare to what actually happens? You might also have students compare a control pan with a pan in which they introduce variables.

Step 3:
After the initial activity has been completed, encourage students to design an additional experiment that can be used to demonstrate convection (see Handout 2). One possibility would be to show the convection currents of smoke trapped in a box. These additional experiments may be performed if time remains in the class period, or as an enrichment activity in a later class period. Groups should describe their experiments, expected results, and conclusions on Handout 2. Students should now be able to diagram a convection cell.

Step 4:
After your students have completed their diagrams on Handout 2, discuss the process of convection. You might draw

a convection cell on the board (see Figure 1), and begin your discussion with the questions on Handout 2.

Help students understand the relationship between convection in water to convection in the mantle, as presented in the "Background Information." Discuss this movement of mantle material as a possible mechanism for movement of the crustal plates.

You might draw on the board a diagram to help explain why scientists think convection is the process that moves crustal plates (see Figure 2).

To check for understanding, have students write a paragraph at the bottom of Handout 2 describing the relationship between their laboratory experiment and plate tectonics.

Enrichment Activities

- ☐ Students can do further research on how scientists discovered that convection might be linked to plate tectonics.
- ☐ Student groups can perform the additional experiments they designed on Handout 2 in a follow-up class period.

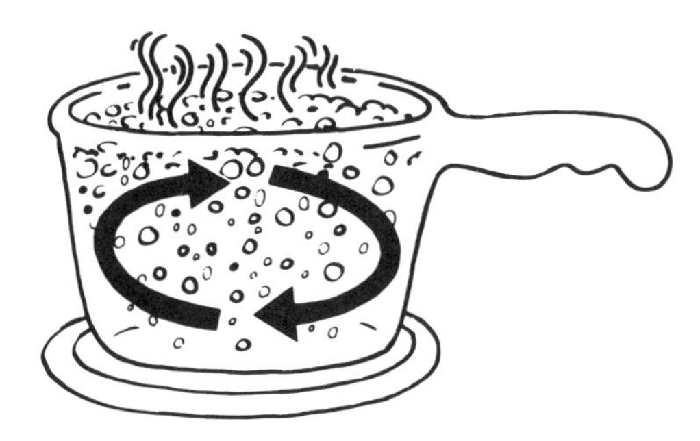

Figure 1. Movement of liquid in a simple convection cell.

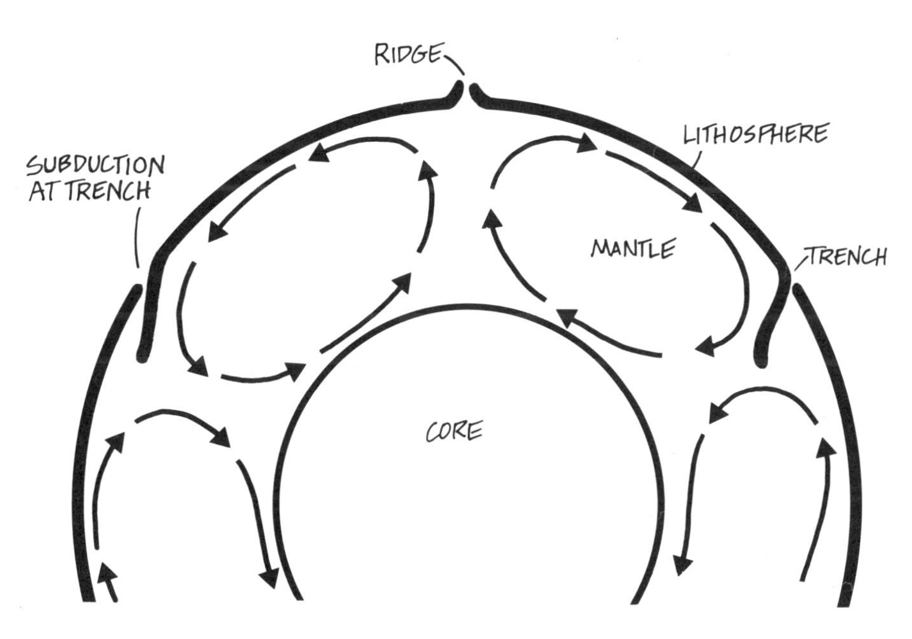

Figure 2. A model of how convection cells within the mantle may be the driving force behind plate tectonics.

Ride the Convection Coaster

What happens when you heat liquids? Find out by doing this lab experiment.

Step 1. Set up the materials for your experiment as shown in the diagram. Your group should have a loaf pan, 2 or 3 beakers, food coloring, an eyedropper, a candle, and matches.

⚠ Always have a beaker of H_2O nearby when using matches.

Step 2. Fill the eyedropper with food coloring. Put the eyedropper in the water, and place a very small drop of food coloring on the *bottom* of the pan. Try not to disturb the water.

Step 3. What happens to the food coloring before you light the candle underneath the pan? Next, light the candle and observe what happens. Try moving your heat source, or think of other variations you can do to make the food coloring move in different ways. Record your observations in the chart below. Use another page if you need more room.

Diagram of set-up	Observations—sketch and/or description

Use What You've Learned

Design another experiment to show what you discovered about convection. Sketch and describe your experiment.

What results would you expect?

From these results, what are your conclusions?

Where do you see this happening every day?

Draw a diagram of a *convection cell.* Indicate the direction of the liquid's movement and the location of the heat source.

How does convection relate to plate tectonics?

A Free Ride on the Earth's Crust

Locating and Measuring Earthquakes

Rob Beattie
Marjorie H. Tobias School
Daly City, California

I began teaching in 1978 after earning my bachelor's degree and elementary education credential from the University of California at Irvine. I taught bilingual first graders for a year, then began my present teaching position in the Jefferson Elementary School District's Gifted and Talented Education (GATE) program.

My partner in the GATE program, Betty Nelson, and I developed a special set of earthquake activities as part of the science curriculum for the second to sixth grade students in this program. These earthquake activities are included in this unit, "A Free Ride on the Earth's Crust." Part of the inspiration for these activities is due to the National Science Foundation-sponsored "Earthquake Institute" that I attended at the University of San Francisco in 1980, and to the instructors of that course: Ray Sullivan, Ray Pestrong, and Herb Strongin.

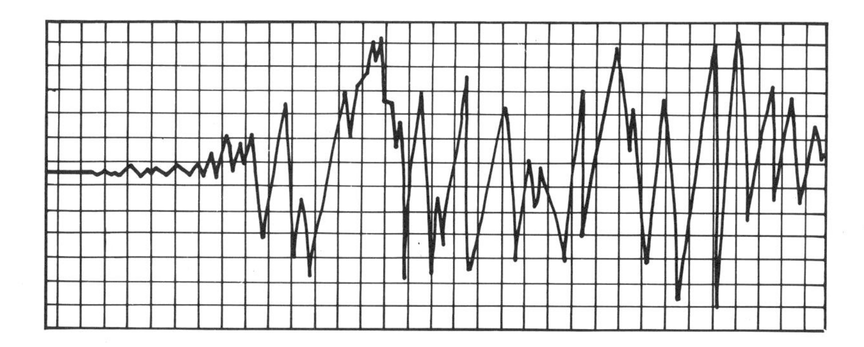

Lessons:

1. "Good Vibrations"
2. "Hang Ten on the Richter Scale"
3. "Surf's Up"
4. "Keep an Eye on the Tide"

Overview

AS A CHILD GROWING UP in southern California, I looked forward to each earthquake with the same anticipation you see on the faces of surfers waiting for the next big wave. The disappointment I felt when I failed to notice a small earthquake was similar to that felt by children who wait in line for a ride on a roller coaster, only to have the ride shut down before they get a turn. Just like the surf, an earthquake is a free ride.

Earthquakes are a part of California life. Although they occur all over the globe, seismic activity is linked in the minds of many people to the Golden State. People living in the Midwest or on the East Coast are not immune to earthquakes, however, and a 5.0 earthquake centered in Kentucky in 1987 was a reminder of that fact. How many of your students know the location of the strongest earthquake to occur in the continental U.S. in the last 200 years? Most students will probably say, "the 1906 San Francisco earthquake." Surprisingly, the answer is an earthquake with an estimated magnitude of 8.7 on the Richter scale centered near New Madrid, Missouri in 1811, which rattled dishes as far away as Boston and Quebec. The area near New Madrid was sparsely populated at the time, and only a few people were killed. However, the earthquake was so severe that it changed the course of the Mississippi River, created a large lake in Tennessee, and was felt over an area of about one million square miles. If the same earthquake were to happen today, much of the nation's midsection would suffer heavy damage.

The earth's crust has been shifting and rumbling since it was formed, and it will continue to shift and change. After long periods of time, tension builds along a fault until the rocks break, releasing the stored tension as waves of energy in an earthquake. These earthquake waves can significantly change the shapes of landforms on the earth's surface in a very short period of time.

We cannot expect the earth to stop moving because we choose to build houses, schools, and other buildings in earthquake zones. Then why do we persist in exposing ourselves to the likelihood of imminent catastrophe? Because severe earthquakes are reasonably rare events, and most of the time do not threaten us. In California, some of us would argue that earthquakes are one of the state's attractions. Not only do earthquakes provide free rides, but they have also played a major role in forming many of California's most scenic areas, such as the coastal and Sierra Nevada mountain ranges. No matter where you live, though, your students will benefit from learning about the reality and inevitability of earthquakes.

Failing to teach students about earthquakes is being shortsighted. Even if your area is more likely to suffer from natural disasters other than earthquakes, there is no guarantee that your state will not be affected by a major earthquake. Moreover, your students may someday live in an area prone to earthquakes, and they should know what to do to protect their safety in the event of a temblor.

Earthquakes are natural phenomena that are catastrophic only when they adversely affect our structures and therefore endanger our safety. The well-being of our students is not served by the typical responses of fear and denial, but rather by careful planning based on informed choice. Fear of the consequences of earthquakes can cause emotional trauma when that fear is nonspecific and unlimited. Knowledge about the behavior of the earth, the likely limits of earthquake shaking, and the resulting damage to our homes and workplaces can help limit fear and its consequences. To deny the danger of an earthquake, especially in areas where the likelihood of a major earthquake is high, is to fail to prepare to survive the experience.

As with any potential natural disaster, our safety lies in adequate preparation. We need to know exactly what is likely to happen to the landforms and buildings around us and then make plans accordingly. In a family, each person must have a job to do, and know what other members' jobs will be. A child with a responsibility during a major earthquake, who knows where all other family members are likely to be and what their jobs are, is a child who is too busy and confident to entertain unreasonable fears or to act out of panic.

The lessons in this unit have successfully led my students to understand the dynamics of seismic movement, and to understand the effects of this movement on their homes and schools. I suggest that you also teach your students about continental drift and plate tectonics, which are the driving forces behind earthquakes (see "Are We Drifting?" on pages 2–41). Armed with knowledge and planning, your students can relax and enjoy the next free ride on the earth's crust.

Key Concepts

- ☐ Earthquakes are natural phenomena that are disastrous only when they affect life or property.
- ☐ Earthquakes produce distinctive wave motions that can be measured and compared to determine the site of their origin.
- ☐ Earthquakes can cause major changes to landforms in a short time.
- ☐ Different landforms respond in distinct ways to seismic movement.
- ☐ Earthquakes and associated landslides, ground subsidence, liquefaction, and seiches can cause significant property damage and personal injury.
- ☐ Properly located, designed, and constructed buildings in earthquake-prone areas can reduce potential damage and injury.
- ☐ Planning, public policy, and individual preparedness are necessary, especially in populated areas, to minimize property damage and injury.

Skills Used in the Lessons

- gathering data
- classifying
- recognizing and predicting patterns
- formulating hypotheses
- problem solving
- understanding space-time relationships
- stating laws
- inventing
- controlling and manipulating variables
- estimating

Extensions and Sources

One logical extension of this unit is family preparedness. Encourage your students to discuss emergency plans with their families. Each family member should know where the others are likely to be and what each is expected to do in case a major earthquake should strike during business/school hours, at night, or on weekends. To help your students make emergency plans, I suggest that you request copies of the free brochure, "Safety and Survival in an Earthquake," from your local chapter of the American Red Cross. Similar information can be found in the front of telephone books in some states, within the section called "Survival Guide."

Earthquake preparedness is more than making emergency plans; it's also checking homes and schools for safety hazards, and then taking steps to reduce or remove those hazards. An excellent book on this topic is *Peace of Mind in Earthquake Country,* by Peter Yanev, Chronicle Books, San Francisco, California, 1974.

If you want to obtain only one easy-to-understand book that contains all you need to know about earthquakes in California and elsewhere, I strongly recommend *Earthquakes: A Primer,* by Bruce A. Bolt, W.H. Freeman and Co., San Francisco, California, 1978. I learned from this book much of the information about the behavior and measurement of earthquake waves contained in this unit. Note that some of the information in this book may differ from the information in your earth science textbook. In researching this unit I found little agreement among textbooks, so I intentionally favored information from geologist-written materials.

Other teaching resources include geologic maps and films. Maps of all states are available from the Public Information Office of the U.S. Geological Survey, 1028 General Services Administration Building, 19th and F Streets NW, Washington, D.C. 20244. Two films that my students profited from are *Continents Adrift—A Study of the Scientific Method,* AMEDFL, 1971; and *Civil Defense—Our Active Earth,* CSDE, 1972.

Students interested in incorporating the study of earthquakes into their careers should know that a variety of professions focus at least partially on this subject, including jobs in geology and the physical sciences, statistics, land use planning, public health and welfare, architectural engineering, foreign relations, data processing, and other technology-related fields.

People who are aware of the implications of life near an active fault must recognize their duty to be personally prepared to be safe and self-sufficient during and after a major earthquake. Students can contact local agencies about plans and provisions for public safety. They can write letters to newspaper editors in order to voice their concerns. Above all, students should know that they cannot blindly trust unspecified bureaucratic agencies for protection during and after this type of natural disaster.

Lesson 1

Good Vibrations
An investigation of seismic waves

Group Size
Individual students

Time Required
2 class periods

Materials
For Day 1:
- transparency, "Types of Seismic Waves," on pages 53–54. Be sure to make transparencies of both pages. Alternatively, make copies of the transparency pages for student handouts.
- one jump rope
- one Slinky toy (Note: This lesson may be taught without the Slinky, by describing the seismic wave movements and relying on your students' prior knowledge of how a Slinky moves. However, I find the lesson works better using the Slinky.)
- roll of masking tape (optional)

For Day 2:
- two 9" by 12" tagboard signs; label one with a large "P," the other with a large "S"
- one small drum and drumstick
- scratch paper and pencils for each student

Key Terms

crust	primary waves
mantle	secondary waves
core	Love waves
seismic	Rayleigh waves
focus	epicenter

Instructional Goals
- To promote an awareness of the role of seismic data in the formulation of theories about the earth's interior.
- To increase understanding of the types and behaviors of seismic waves.

Student Objectives
Students will:
- Demonstrate in class discussion an awareness of the types and behaviors of seismic waves.
- Participate in simulations of wave movement and detection.
- Calculate the ratio of travel speeds of seismic waves.
- Discuss the conclusions inferred from the relative speeds of primary and secondary seismic waves.

Prerequisite Knowledge
Students should understand the basic concept of plate tectonics—that the earth's crust consists of several large slabs that slowly move on top of a layer of molten rock. Students should also be capable of reducing fractions.

Advance Preparation Time
About 15 minutes each day
For Day 1:
- Review the "Background Information" and "Lesson Procedure."
- Gather materials.
- Make the overhead transparencies, or duplicate these pages as handouts.
- Take a few minutes to practice the lesson procedure with the Slinky.

For Day 2:
- Gather materials.
- Make the two small signs labeled "P" and "S."
- Practice the demonstration with your assistants.

Teacher Tip
- Attach small pieces of masking tape onto each of the Slinky's coils in a row along one side, making the travel of the waves down its length more apparent.

Background Information

The earth's crust consists of several large and fairly stable slabs called plates. Each plate extends to a depth of about 80 kilometers. The plates slowly move horizontally, relative to each other, on top of a layer of softer molten rock. At the edges of the plates, large deforming forces affect the rocks, causing physical and chemical changes in them. It is at these plate edges that the greatest changes occur to the earth's geological structure. In a nutshell, this describes the process of plate tectonics (see the unit "Are We Drifting?" on pages 2–41).

As the plates move, they sometimes produce earthquakes. The shock waves we feel when an earthquake strikes are called seismic waves. Just as surfers classify ocean waves by their behavior and appearance, geologists recognize four distinct types of seismic waves by the way they move. (See the transparency, "Types of Seismic Waves," on pages 53–54 for illustrations.)

Two of these types of seismic waves pass through the body of the earth and thus are called *body waves*. Body waves originate at an earthquake's *focus*, the place deep within the earth's crust or upper mantle where the actual movement of rock has taken place. Most movement on the well-known San Andreas Fault along the West Coast of the U.S. occurs three to seven miles below the surface. By studying the behavior of body waves as they pass through the earth, scientists have deduced much of what we know about the structure of the earth's interior.

The first type of body wave is called the *primary wave* (or P-wave). The P-wave is like a low-frequency sound wave. It is conducted through solids, liquids, and gases. Sometimes P-waves can be heard by the human ear as a low rumble. P-waves (also called pressure waves) travel at approximately 5.5 kilometers per second through granite, and can be measured by seismographs over most of the earth's surface. A P-wave feels like a jolt and generally causes little damage.

The second type of body wave is called the *secondary wave* (or S-wave), which travels through granite at approximately 3 kilometers per second. S-waves travel only through solids, and cannot travel through liquids. This inability to be conducted through liquids leads scientists to believe that the earth's outer core is liquid, because S-waves are measured by seismic stations on only those parts of the earth's surface that are not within the 143-degree angle blocked by the earth's core (see Figure 1 on page 48). The transverse wave motion of S-waves will not pass through liquids or gases because in these less-dense forms of matter some molecules are moved only from side to side, not forward. Thus, the S-waves do not push forward through liquids as the P-waves do. The inaudible S-wave is sometimes called a shear wave because it moves the earth's surface sharply from side to side. S-waves cause more damage to man-made structures than P-waves do.

The remaining two types of seismic waves are *surface waves* that do not pass through the earth's interior. Surface waves originate at the earthquake's *epicenter*, the place on the earth's surface that is directly about the focus (see Figure 2). The *Love wave* (or L-wave) causes the most damage to structures because it shakes the surface back and forth at right angles to the direction of its travel, horizontally shaking the foundations of buildings.

The *Rayleigh wave* is the slowest seismic wave. It rolls along like an ocean wave, both horizontally and vertically in the direction of the wave's travel. If you could ride a Rayleigh wave on a surfboard, the actual path of your movement would be an ellipse.

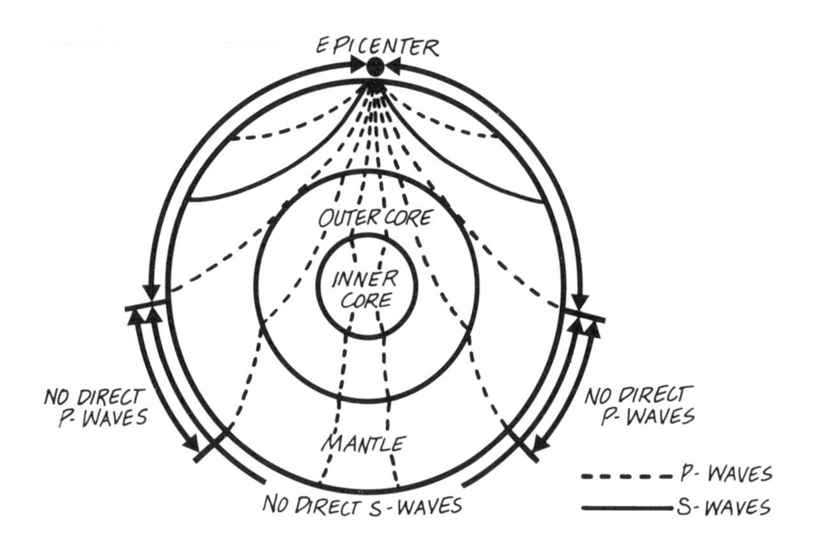

Figure 1. Cross-section of the earth's interior showing the paths of P-waves and S-waves as they travel outward from an earthquake's focus. Notice the zone that shows no direct S-waves, indicating that the outer core is liquid.

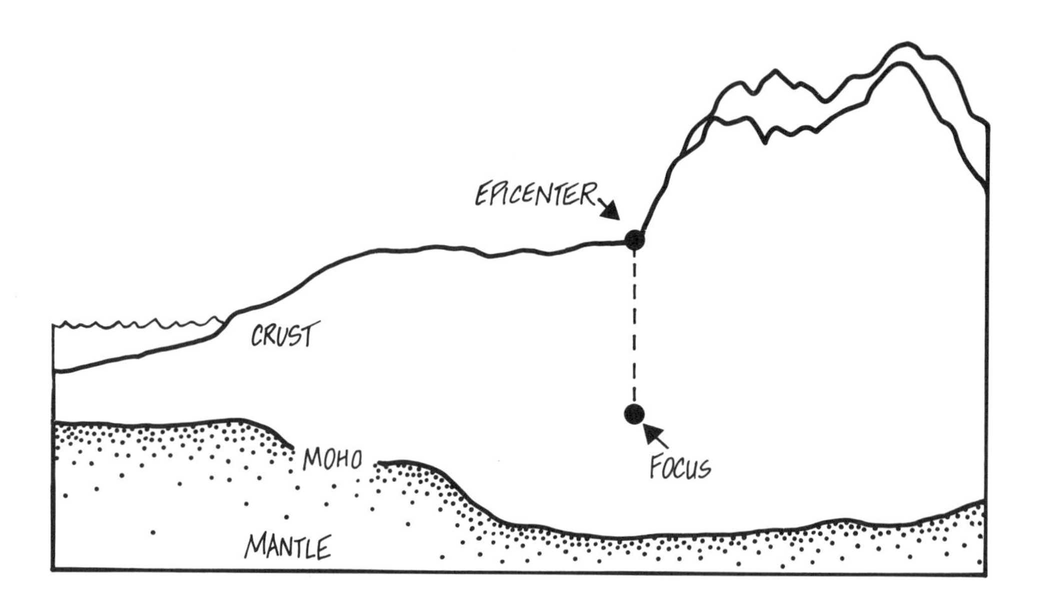

Figure 2. A cross-section of the crust showing the focus and epicenter of an earthquake.

Lesson Procedure

For Day 1

Step 1:
Review the earth's interior structure with your students. Draw a diagram like Figure 1 (page 48) on the chalkboard. For the sake of simplicity, only the three main parts of the earth's interior are shown here. Actually, the reflection and refraction of seismic waves as they pass through the earth's layers lead scientists to believe that the mantle and core are each composed of layers that vary in density and rigidity.

Step 2:
Discuss and review information about plate tectonics and the constant motion of the earth's crust, as presented in "Background Information." Elicit from students examples of evidence that indicate the earth's surface is shifting; for example, volcanoes, earthquakes, faults, folded strata, landslides, etc.

Draw a diagram similar to Figure 2 (page 48) on the chalkboard. Be sure to explain the difference between an earthquake *focus*, the section of the fault underneath the earth's surface that has moved, and its *epicenter*, the area on the earth's surface directly above the focus.

Step 3:
Display the overhead transparency, "Types of Seismic Waves," from pages 53–54. Reveal the transparency one section at a time as you demonstrate the different wave types in Steps 4-6. This helps students to focus on the relevant information. Students should take notes from this transparency, or you might duplicate the transparency pages as handouts. Use material from "Background Information" as needed.

Describe primary waves, referring to the diagram and information on the transparency. To demonstrate P-waves, place a Slinky toy on a tabletop in view of the entire class. Ask a volunteer to grasp one end of the Slinky while you control the other end. With the Slinky stretched between you, squeeze 10-15 coils tightly together, then release the coils suddenly. The wave will travel the length of the Slinky and then will bounce back (see Figure 3).

Discuss your students' observations. This forward push-pull motion of the P-waves conducts energy forward, and is similar to the action of sound waves that push through the air to affect your ears. If necessary, point out that the wave is reflected back along the Slinky. Ask your students what a reflected sound wave is called. (Answer: an echo.) In fact, the echoing of P-waves off the crust and back into the earth can double the surface disturbances in an earthquake. This explains why miners working deep below the surface have reported less shaking during some earthquakes than that observed on the surface.

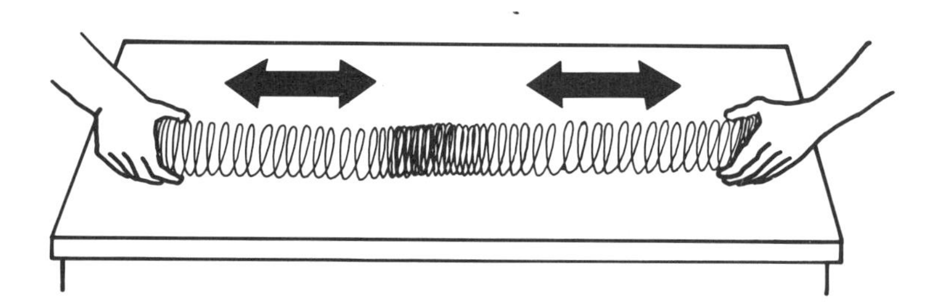

Figure 3. Demonstrating P-waves using a Slinky.

Step 4:
Describe secondary waves, referring to the diagram and information on the transparency. To demonstrate S-waves, use the jump rope on a tabletop, another volunteer, and the same procedure as in Step 3. This time the motion is achieved by whipping your end of the jump rope back and forth while the volunteer holds the other end still, slightly beyond the edge of the table (see Figure 4).

Remind students that S-waves begin at the earthquake's focus and then whip out toward the surface, so that they are felt as a back-and-forth shaking motion on the surface. As the source of the wave in this demonstration, you represent the focus of the earthquake. Your assistant is holding the other end of the rope just beyond the end of the table, so that the table edge represents the earth's surface. The movement that a person on the earth's surface might feel is like the movement of the rope back and forth at the table's edge, similar to a whip crack.

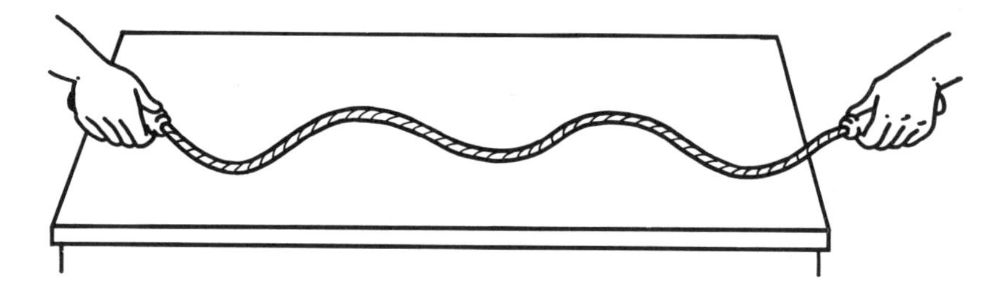

Figure 4. Demonstrating S-waves using a jump rope.

Step 5:
Describe Love waves, referring to the diagram and information on the transparency. To demonstrate L-waves, use the Slinky and another volunteer. Place the Slinky on a tabletop. Take your end of the Slinky and whip it back and forth (see Figure 5), just as you did with the jump rope in Step 4.

The difference between L-waves and S-waves is more apparent when you recall that the S-waves start at the earthquake's focus and end at the surface. Thus, we feel only the "whip crack" part of the movement. The L-wave, on the other hand, travels *along* the earth's surface, displacing rocks and soil from side to side, which can shift foundations beneath buildings. The deeper the earthquake's focus beneath the surface, the less displacement from L-waves occurs on the surface. The only way that L-waves affect bodies of water is by shaking their banks.

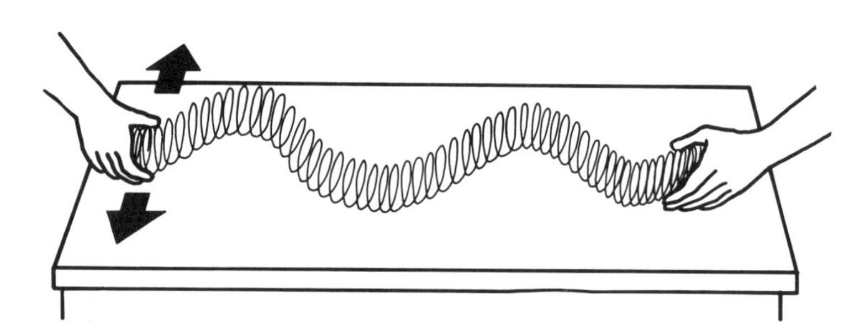

Figure 5. Demonstrating Love waves using a Slinky.

Step 6:
Describe Rayleigh waves, referring to the diagram and information on the transparency. To demonstrate Rayleigh waves, place the Slinky on a tabletop and hold one end of the Slinky in each hand. Alternately flex your wrists up and down so that an ocean-like wave rolls along the Slinky (see Figure 6). This may take a little practice.

Explain that the Rayleigh wave is like an ocean wave that rolls through the earth, slowly moving the surface up and down. A rock moved by these waves travels in an elliptical path. A piece of masking tape on the side of one of the coils will neatly demonstrate this elliptical motion. Because Rayleigh waves move up and down, they also affect bodies of water.

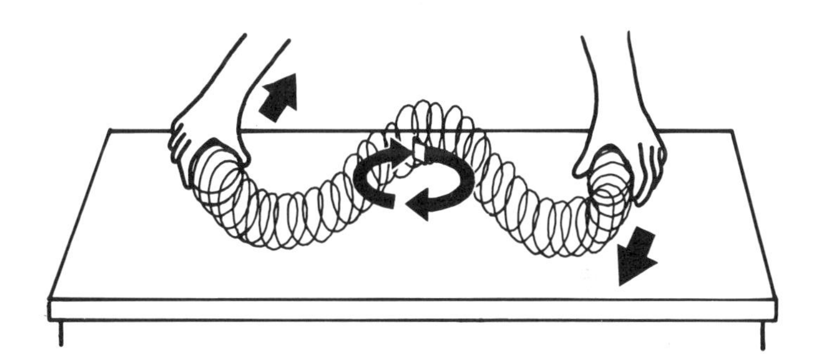

Figure 6. Demonstrating Rayleigh waves. Notice the elliptical path of the masking tape marker on the Slinky.

For Day 2
Step 1:
Review the four types of seismic waves and their characteristics as presented in the previous class period, emphasizing the different traveling speeds of P-waves and S-waves.

Step 2:
Introduce the activity, in which your students will record the traveling speeds of simulated P-waves and S-waves and then calculate the P-wave:S-wave ratio. To do the activity, arrange your students in a long, straight line in a corridor, along a fence, or on the school's playing field. Each student should have a pencil and paper for recording information.

Together with an assistant, stand at the beginning of the line. In one hand, hold the sign with the large "S" written on it, which stands for the S-wave; your assistant should hold the sign marked "P" for the P-wave. Another volunteer should start to rhythmically beat a drum at the rate of one beat per second (time this with the second hand of a wrist-watch).

Step 3:
At the same time, which represents the occurrence of an earthquake, the P-wave and S-wave (you and your assistant) start to walk along the line of students. Count out loud and in unison the number of beats so students can hear the number of seconds that have elapsed as each wave passes in front of them. The P-wave (your assistant) must take one step per second, the S-wave (you) must take one step every two seconds. This rate approximates the 3:5.5 ratio of the P-wave and S-wave traveling times closely enough for the sake of this demonstration. Students should write down the time in seconds as each wave walks past them.

When both waves (you and your assistant) have completed the trip, everyone returns to the classroom. If necessary, do the activity more than once to make sure each student has recorded the times.

(Note: You'll find this demonstration much easier if you practice pacing and counting with your assistant. The drummer helps the two demonstrators keep in step, which can become increasingly difficult as the distance between the P-wave and S-wave grows larger. You and your assistant need to take steps of equal length throughout the simulation, or else the students' calculations won't work. If you find it works better, you might have your assistant be the S-wave instead of the P-wave.)

Step 4:
When back in class, ask each student to write the arrival times of the waves as a fraction: P/S. Students should reduce the fraction to its simplest terms.

Ask students to share their answers. You'll find that student answers may vary, but if care has been taken with the counting and pacing during the demonstration, the answers should all approximate 1/2. Collect enough answers (data) until your students discover this approximation, or point out that each student's result is close to 1/2. Elicit from your students the conclusion that the P-waves move about twice as fast as the S-waves.

Demonstrate that this fraction can be expressed as a ratio, 1:2 (spoken as "one to two" or "one for every two"). This ratio describes the constant relationship between the traveling times of the waves. The P-wave takes one-half as many seconds as the S-wave to cover a given distance. Put another way, for every kilometer that an S-wave travels, a P-wave travels 2 kilometers.

Step 5:
To wrap up the lesson, ask your students how this knowledge of P-wave and S-wave traveling times might be useful to seismologists. Accept all responses, but delay providing the answer—that scientists use this lag time between the arrival of P-waves and S-waves to determine the origin of an earthquake—until the next lesson. This will keep students thinking about the question.

Step 6:
To lead in to the next lesson in this unit, assign the following homework project. Challenge each student to build a seismograph at home, which should be ready for presentation at the next class meeting. Do not describe how a seismograph works or what it looks like, but define it only as a device that records movement on the earth's surface.

When assigning this project, I find that this intentional ambiguity in the directions results in more creativity on the students' part. With too much description, students tend to build frames with pencils hanging from them, while students who have no preconceptions are more apt to make amazing contraptions, perhaps with marbles that land on tin pans when the device is shaken, or electric switches that ring bells or stop clocks. You'll find that students who complain the most about your vague directions sometimes come up with the most clever solutions.

Enrichment Activities

- ☐ Ask students what would happen to the ripples of water in a basin if you splashed the water at one end. Would the ripples travel the length of the basin? Would they bounce off the other end and ripple back? Would the ripples bounce back if the basin were twisted or jerked sideways? This failure to remain an intact wave and bounce back is the characteristic that prevents S-waves from passing through liquids.
- ☐ Assign one or more students to design experiments to investigate the motion of waves in water. Students should make conclusions about wave motion as related to seismic waves.

Types of Seismic Waves

Body waves originate at the earthquake's **focus** (deep within the earth).

1. Primary waves (P-waves)
- Speed: 5.5 km/second (fastest)
- Arrival: first
- Damage: felt as a jolt, little damage
- Medium: travels through solids, liquids, and gases
- Movement: push-pull

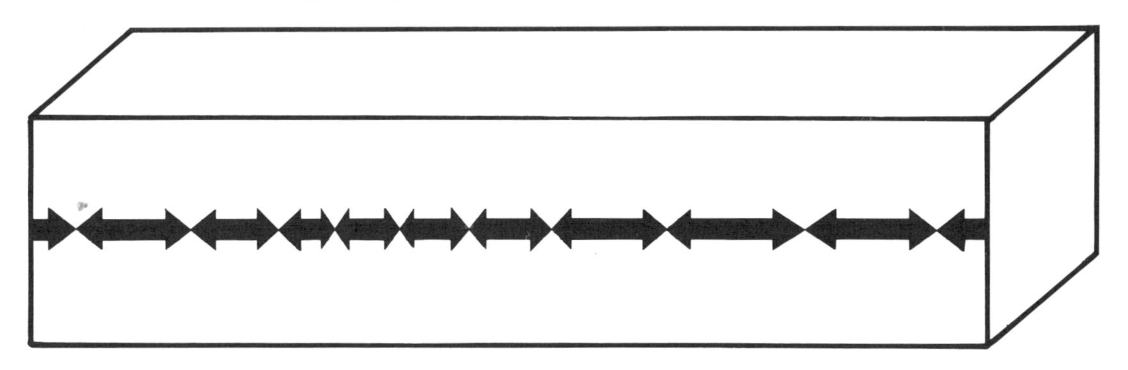

2. Secondary waves (S-waves)
- Speed: 3 km/second
- Arrival: second, after P-waves
- Damage: jerking back and forth can cause significant damage
- Medium: travels through solids only
- Movement: whiplike, at right angles to direction of travel from focus to surface

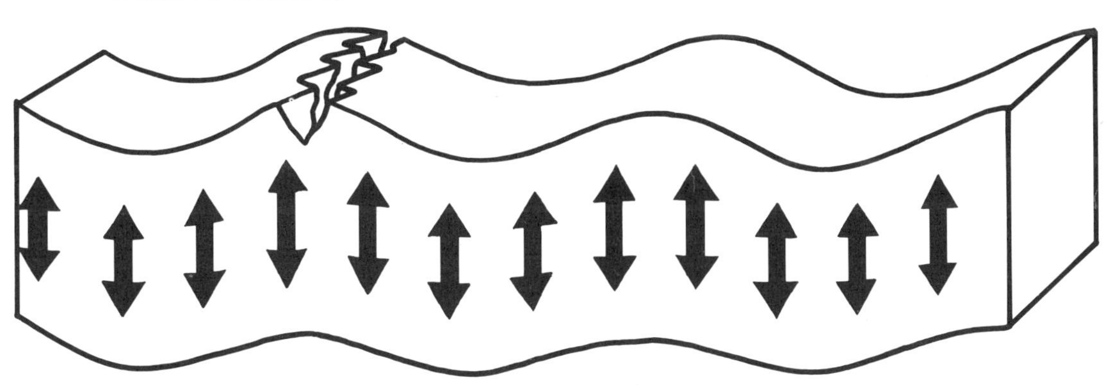

Surface waves originate at the earthquake's **epicenter** (on earth's surface above the focus)

1. Love waves (L-waves)

- Speed: slower than S-wave
- Arrival: third, after S-waves and P-waves
- Damage: buildings shift on their foundations as ground tosses from side to side
- Medium: travels through solids only
- Movement: tossing, at right angles to direction of travel over surface; intensity weakens deeper below the surface

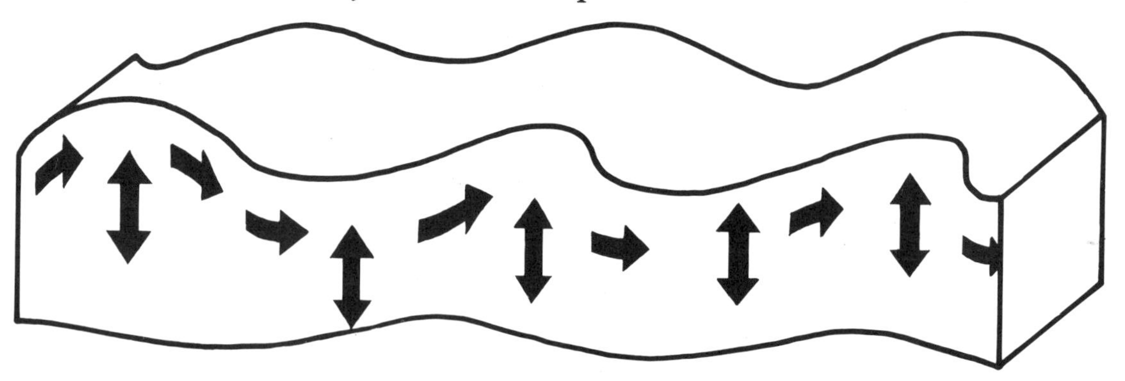

2. Rayleigh waves

- Speed: slowest
- Arrival: last
- Damage: tilting of structures causes some damage
- Medium: travels through solids and liquids
- Movement: rolling up and down, back and forth

Notice the elliptical path, similar to an ocean wave.

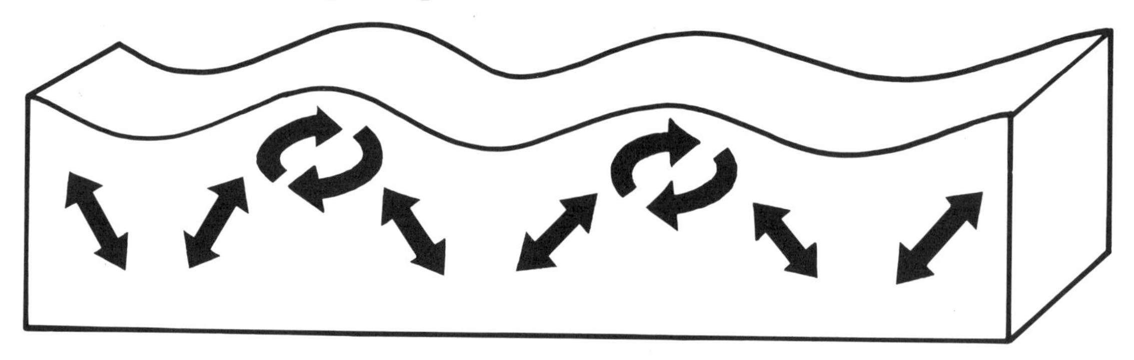

Hang Ten on the Richter Scale

Designing seismographs and locating earthquakes

Group Size
Individual students and small groups

Time Required
2 class periods

Materials
- each student's home-made seismograph (see Step 6 on page 52 for instructions on this assignment)
- copy for each student of Handout 1, "Find the Epicenter," on pages 61–62
- one drafting compass for each student

Key Terms
seismograph
lag time
focus
epicenter

Instructional Goal
- To demonstrate the scientific methods of locating and measuring earthquakes.

Student Objectives
Students will:
- Develop criteria for evaluating their home-made seismographs, then test and evaluate their seismographs.
- Use the information about seismographs obtained in the evaluations to design improved seismographs.
- Use data and formulas provided on Handout 1 to locate the epicenter of a hypothetical earthquake.
- Demonstrate in class discussion and in the activity an awareness of the methods that seismologists use to locate the epicenters of earthquakes.

Prerequisite Knowledge
Students should be able to distinguish the four types of seismic waves, as presented in Lesson 1. In addition, students should understand the principle of inertia: that objects tend to remain at rest unless acted on by an outside force. Students should also know how to draw circles using a drafting compass.

Advance Preparation Time
About 15 minutes
- Review the "Background Information" and "Lesson Procedure."
- Gather materials.
- Make copies of the student handout.

Teacher Tips
- In evaluating the home-made seismographs, I like to have my students vote to select the criteria they think are the most appropriate.
- Delegate seismograph testing to one or more students.

Background Information

When you sit on an ocean beach, you can feel the force of the crashing waves as their energy vibrates through the sand. In the same way, seismographs can sense the vibrations of seismic waves that originate in faraway earthquakes.

A typical seismograph design includes a base and frame that are firmly attached to bedrock. These parts will vibrate and move when the earth does. A rotating cylindrical drum is attached to the seismograph's base. The drum will also move with the earth and the rest of the apparatus.

A heavy weight is suspended from the frame, hanging on a flexible spring or wire. Since most of an earthquake's vibrations are absorbed by the spring or wire, the weight remains stationary, suspended in space during an earthquake. (This illustrates the principle of inertia.) A stylus is attached to the bottom of the weight, which draws a line on the rotating drum (see Figure 1). This line is synchronized to local time, and it is normally a straight line that circumscribes the rotating drum. During an earthquake, however, the drum moves from side to side, and the suspended stylus draws a zigzag line on the drum.

Each of the four types of seismic waves leaves a distinctive zigzag line on the seismograph recording, so the exact time of each wave type is accurately recorded (see Figure 2 on page 57). The difference in the arrival times of the various waves, and our knowledge of seismic wave speeds, allows us to accurately locate the earthquake's epicenter.

Seismologists use a process called *triangulation* to pinpoint the epicenter of an earthquake. This process involves, as its name implies, recordings from at least three different seismograph stations. When an earthquake is recorded on a seismograph, the seismologist first distinguishes between the zigzag lines that indicate P-waves and those that indicate S-waves. The seismologist notes the arrival times of the P-waves and S-waves, then calculates the *lag time*, or difference between the two

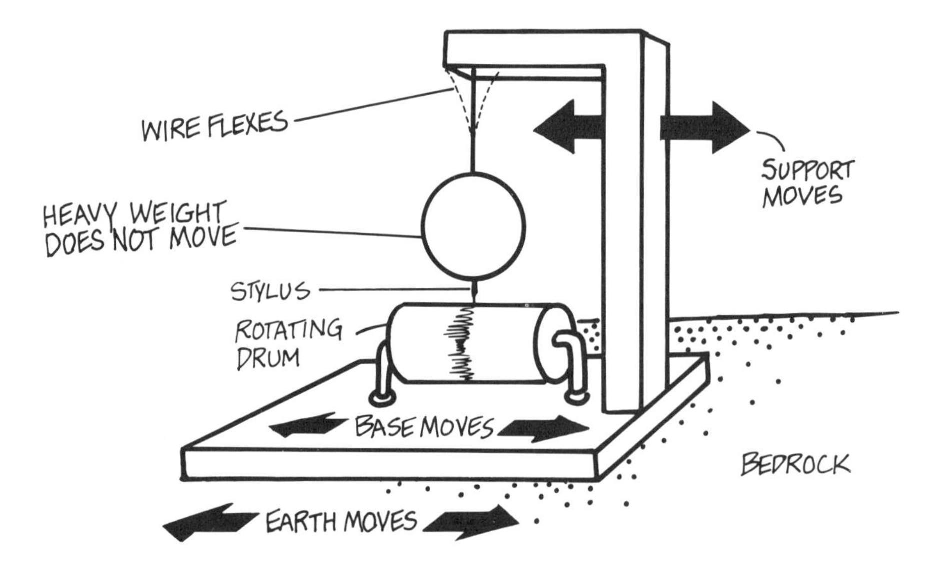

Figure 1. Components of a typical seismograph. The arrows indicate the parts of the seismograph that move during an earthquake. Note that the stylus remains still while the base and rotating drum move, producing the familiar zigzag recording of an earthquake.

arrival times. Since the traveling speeds of P-waves and S-waves are known, the seismologist can then calculate how far the seismograph is from the focus, which is approximately the same distance to the epicenter.

This distance, however, does not tell which direction the epicenter is from the seismograph—the epicenter could be located anywhere along a circle with a radius of the calculated distance. Thus, data from three seismograph stations is needed to determine the epicenter (see Figure 3).

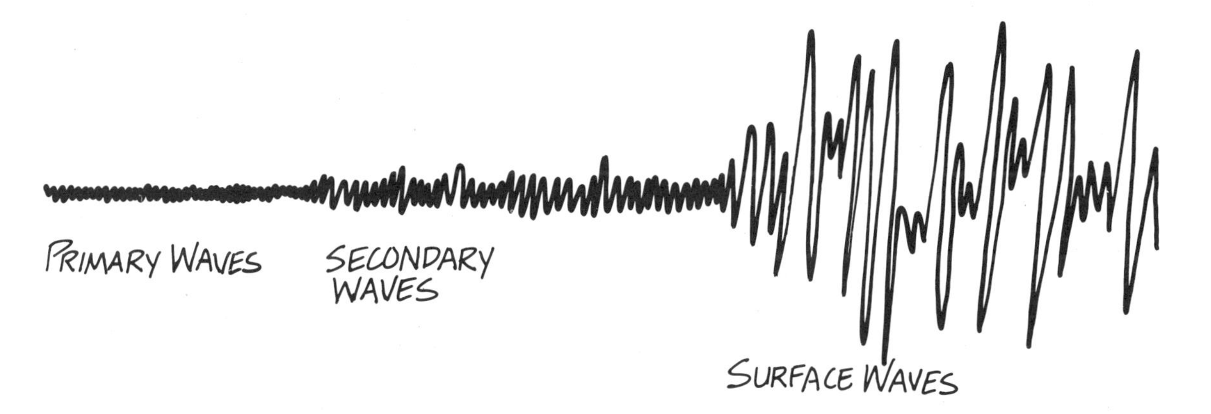

Figure 2. A seismograph recording showing the arrival times of the different types of seismic waves.

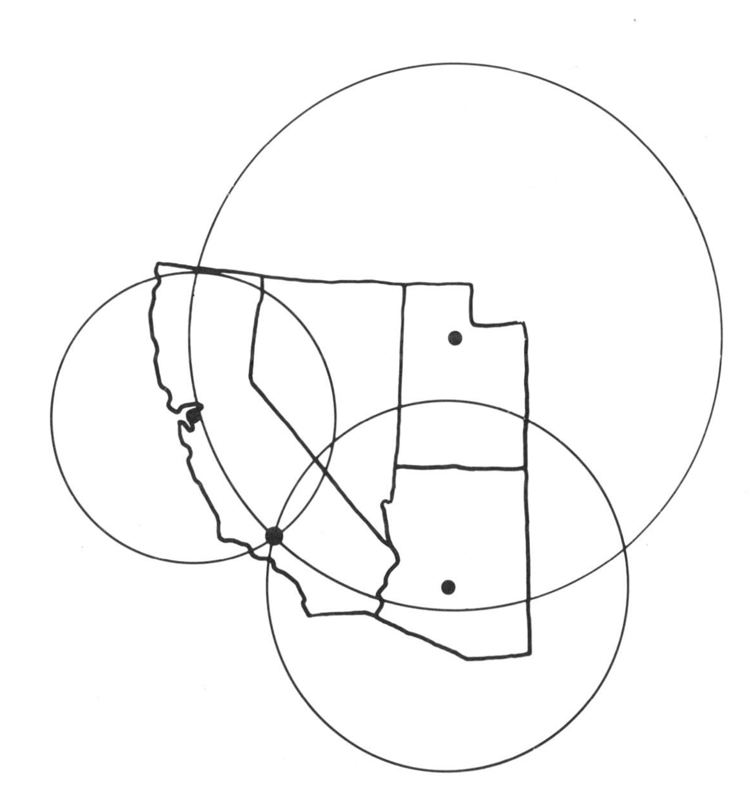

Figure 3. Triangulation used to determine an earthquake's epicenter.

Lesson Procedure

For Day 1

Step 1:
Review Step 6 on page 52 in the previous lesson for instructions about the seismograph homework assignment. Give these instructions to your students prior to doing Lesson 2.

Step 2:
Invite students to propose a test that includes a set of criteria to fairly evaluate the efficacy of all the seismographs. For example, one test might be to place all the seismographs on the same table, then hit the table with a series of blows of increasing intensity, noting at which intensity each seismograph responded. Or, students might place greater value on seismographs that leave permanent or continuous records, or those that respond differently to jolts of varying force.

Write a list of the proposed tests on the chalkboard. When students have proposed all the tests they can think of, have the class select the most appropriate one(s), perhaps by assigning weights or priorities to their tests.

Step 3:
Proceed to test each of the students' seismographs. You might assign a student or teams of students to perform the tests.

Ask students to look for common features of the successful models, as well as common problems of the less successful designs. Record these features and problems in two columns on the chalkboard.

Another method of evaluation is to have students prepare sheets of notebook paper with their names on top and four columns: "How well it works," "Neatness," "Originality," and "Total points." Then have students rate each seismograph design on a scale of 1 to 10 (10 being the best) in each of the categories. It's not necessary to emphasize a winner, but a fair assessment by peers is an invaluable teaching tool. Use this time to write down your own evaluation of student contraptions.

Step 4:
Divide the class into small groups. Ask each group to sketch or design a seismograph that incorporates the key features of the successful models while avoiding the problems of the less successful instruments. Discuss their designs and how well these designs meet the criteria for success.

Step 5:
Explain how an actual seismograph works, as described in the "Background Information." You might make copies of Figure 1 on page 56, or draw it on the chalkboard.

For Day 2

Step 1:
Review seismic wave motion simulations as done in Lesson 1, emphasizing the difference between the four types of waves. Ask students to describe the relationship between the speeds of P-waves and S-waves.

Step 2:
This part of the lesson teaches students how to use triangulation to determine the epicenter of an earthquake.

Draw a dot on the chalkboard and label it "A," representing seismograph station A. Tell your students that the focus of a hypothetical earthquake has been determined to be 30 kilometers away from station A. With this information, ask students to pinpoint the earthquake's location. (Use 1 cm = 1 km as a convenient scale on the chalkboard.) Accept all student responses, until someone suggests that the earthquake could have occurred anywhere along a *circle* with a radius of 30 kilometers from station A.

Draw a circle around station A on the chalkboard to represent all the locations on the earth's *surface* that are 30 km from station A. Ask students how they might narrow down the possible locations to one. Their suggestions might include talking to people who live near the earthquake to find the relative intensities of shaking in different locations, or visiting several locations to look for physical evidence of damage, etc.

Step 3:
If not suggested by a student, add to the discussion the possibility of using readings from another station. Draw another dot on the chalkboard, and label this dot "B."

Tell students that the distance of station B from the focus of this hypothetical earthquake has been determined. Draw a circle with station B as its center that intersects the circle around station A in two places. Ask students to indicate the possible locations of the epicenter. There are two possible locations—the two places where the circles intersect. These are the only two places that are 30 kilometers from station A and also the determined distance from station B. Ask students for suggestions to determine which point is the epicenter.

Step 4:
If not suggested by a student, add the possibility of using readings from a third seismograph station. Draw a third dot on the board and label it "C." Draw a circle around it that intersects the other two circles at one of the two points (see Figure 4).

Ask a student to mark an "X" on the board to indicate the epicenter. Tell your students that this method is called *triangulation,* and it is used by seismologists to locate the epicenters of

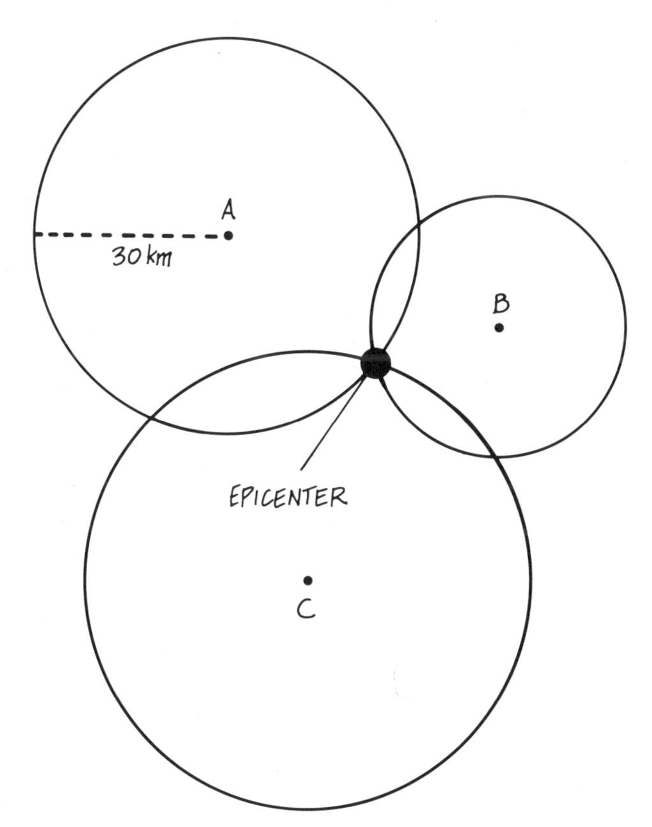

Figure 4. Example of triangulation illustration on chalkboard.

earthquakes all over the world, even in the middle of oceans.

It is important to note that the source of the earthquake waves, the focus, is below the earth's surface. Seismographs are measuring the distance of the recording station to the focus, not the epicenter. Therefore, the true epicenter of an earthquake is actually somewhere *near* the place that the three circles intersect, not exactly that distance from the three reporting stations.

Step 5:
Distribute Handout 1, "Find the Epicenter," and a compass to each student. If necessary, review with your students how to draw circles with a compass.

Have students read the instructions on the handout silently while you read them aloud. Answer any questions they may have about calculating the lag times. Students should complete the chart by calculating and recording the difference and distance for each seismic station. Make sure that the students all have approximately the same answers (see the Answer Key below).

Have students complete the handout by using their compasses to draw circles around the three dots that have as their radii each station's distance from the earthquake. Students have already calculated and written these distances on their handouts.

When students are finished, they should have three circles with points A, B, and C as their centers. These three circles should intersect at one point. Each student should mark this point, the epicenter of the earthquake, with an "X."

As a check, ask students where, if anywhere, a fourth station's circle would intersect the other three circles. Of course, every circle would intersect the others at the same point.

Enrichment Activities

- ☐ Arrange a trip to a local university's geology department or a local office of the U.S. Geological Survey to see an operating seismograph and talk to experts. Seismograph recordings of major or recent earthquakes are typically available for viewing at these locations.
- ☐ My students love to debate. When evaluating the homemade seismographs, perhaps two or more of your students who have a strong advocacy for their own or another's seismograph design might debate about their design's superiority.
- ☐ Invite a local geologist to evaluate your students' seismographs. Or talk to your local newspaper about covering a "Seismograph Testing Day" at your school.

Answers to Handout 1.

	Arrival Times			Distance from
Station	**S-wave**	**P-wave**	**Lag Time (S – P)**	**Epicenter**
A	12:00:18.5	12:00:11	7.5 sec	65 km
B	12:00:12	12:00:7.25	4.75 sec	40 km
C	12:00:44	12:00:27	17 sec	150 km

(Note that the scale on the handout, which is used for determining the distance from the epicenter, does not have division marks for every kilometer. Thus, your students' answers in this column may vary slightly from the answers given.)

Find the Epicenter

To locate an earthquake's epicenter, we need seismograph readings from three different stations. For each station, compare the arrival times of the P-waves and S-waves, then calculate the difference in seconds, which is called the "lag time." (S-wave arrival time – P-wave arrival time = lag time) Record the lag time for each station on this chart.

	Arrival Times			
Station	S-wave	P-wave	Lag Time (S – P)	Distance from Epicenter
A	12:00:18.5	12:00:11	________ sec	________ km
B	12:00:12	12:00:7.25	________ sec	________ km
C	12:00:44	12:00:27	________ sec	________ km

Next, use the scale below to determine how far each station is from the epicenter. To do this, find the lag time in seconds on the right side of the scale. On the left side of the scale, estimate the distance. For example, a lag time of 2 seconds would mean a distance of about 12 km. Record each estimated distance on your chart, then go on to page 2.

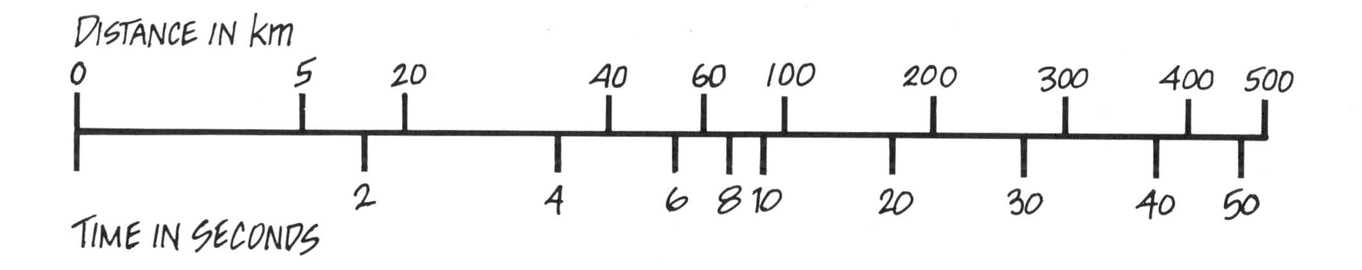

If you have finished filling in the chart on page 1, you are now ready to find the earthquake's epicenter in the same way a seismologist does. Use a scale of 1 mm = 1 km to draw a partial circle around station A below with a radius of the "Distance from Epicenter" from your chart. Repeat for station B and station C. Mark an "X" on the epicenter (where the three circles intersect).

A

B

C

Lesson 3

Surf's Up

Simulating geologic hazards

Group Size
Individual students and whole class

Time Required
1 or 2 class periods

Materials

- one child's plastic sandbox, or a large cardboard box (at least 30 inches by 30 inches by 6 inches) lined with plastic; half-filled with sand or soil
- two yardsticks, or two 1-inch by 2-inch (or 1-inch by 4-inch) boards. These should be set inside the sandbox near the bottom, with the ends of the yardsticks extending through holes made in the sides of the sandbox, providing handles (see Figure 2 on page 67).
- materials to build a model community: toy houses or blocks, toy trees or twigs, pipe cleaners, plastic plants, straws, string, etc.
- one bucket filled first with sand and then with water. The level of the water should be about 2 inches below the surface of the sand.
- one toy house or block
- transparency, "Landforms and Architecture," on pages 70–71. Be sure to make transparencies of both pages. Alternatively, make copies of the transparency pages for student handouts.

Key Terms

bedrock	foundation
landfill	tsunami
alluvium	seiche
mudflat	reservoir
liquefaction	

Instructional Goals

- To demonstrate the effects of a major earthquake on various landforms and types of buildings by simulating an earthquake in a sandbox and by simulating liquefaction.
- To enable students to recognize the relationship between a building site's underlying geologic structure and the earthquake-resistant architecture best suited for the site.
- To promote student understanding about building codes designed to protect structures and people in the event of a major earthquake.

Student Objectives
Students will:

- Prepare a model community in a sandbox and hypothesize about the effects of simulated earthquake shaking on the structures and landforms they have built.
- Evaluate the results of the simulated earthquake and generalize about the effects of seismic wave motion on various types of landforms and structures.
- Generalize about the possible effects of seismic activity on local landforms and structures.

Prerequisite Knowledge
Students should be familiar with the various types of seismic waves and the ground shaking they produce at the earth's surface (from Lesson 1).

Advance Preparation Time
About 30 to 60 minutes

- Review the "Background Information" and "Lesson Procedure."
- Gather materials.
- Make the transparencies or student handouts.
- Prepare the sandbox and bucket as described in "Materials."

Teacher Tips

- Practice the sandbox earthquake ahead of time. Hold one yardstick on one side of the sandbox and have an assistant hold the other yardstick on the other side of the sandbox. On a count of 3, sharply pull on the yardsticks, giving the sandbox a strong jolt (see Figure 3 on page 67). If you have produced a jagged fault-line in the sandbox, and the surface has had a good shake, then your sandbox is ready for the class activity.
- Wrap rubber bands around the ends of the yardsticks to provide resistance to movement. Some source of friction between the yardsticks (sandpaper, for example) will ensure better results.
- Make sure students don't jiggle the bucket filled with sand and water before the activity, or it will liquefy prematurely.
- Find photos in earth science texts or resource books of surface changes resulting from an earthquake (cracks, landslides, etc.) and earthquake-damaged buildings. These photos can be used to reinforce your students' hypotheses and experiment results. A particularly good book with these types of photos is *Peace of Mind in Earthquake Country*, by Peter Yanev, Chronicle Books, San Francisco, California, 1974.

Background Information

Generally speaking, the success or failure of a building to safely "ride out" earthquake waves is determined by three variables: the landform(s) under or near the structure, the structure's distance from the fault or epicenter, and the architecture and materials of the structure. The purpose of this lesson is to demonstrate to your students the importance of these variables in a simulated earthquake.

Cities in several earthquake-prone states, such as California and Alaska, require developers to adhere to strict building codes that specify materials and construction practices that result in safer structures. Ordinances also require that earthquake and geologic information be made available to prospective homeowners. The areas of greatest risk are densely populated urban centers located near active faults, although earthquakes occur all over the earth's crust and some of the worst earthquakes in recorded history have struck in areas not considered earthquake-prone. In fact, geologists believe that the lack of regular seismic activity predisposes areas near faults to more severe shocks than areas that experience continual "creep," or small periodic earthquakes that release tension.

To reduce the risk of damage in a major earthquake, consumers and builders alike must realize the basic principles of how earthquakes affect structures. More damage occurs to buildings closer to the earthquake's epicenter than to similar buildings on similar sites farther away from the epicenter. Buildings that rest on stable geologic formations closer to the earthquake's epicenter may survive with less damage than buildings built on unstable geologic formations that are farther away from the epicenter. Additionally, certain building materials and types of structures suffer less damage than others and thus are preferable for use in fault zones.

The importance of building on stable geologic formations cannot be overemphasized. It is best to build on bedrock, because the denser and stronger the rock, the more earthquake shock is absorbed. Dense molecules provide excellent conduction of seismic waves, so the waves move through the medium quickly and little shaking occurs. Unconsolidated rocks and soil, on the other hand, tend to shift around when shaken, resulting in tilted or collapsed buildings. On slopes, landslides may endanger even those buildings constructed on bedrock outcrops.

Two examples of particularly unstable geologic formations are landfill and mudflats, such as those that surround San Francisco Bay in California. These landforms will be subject to a great deal of earthquake shaking because they are not consolidated at all. Seismic waves are slowed by the low density of soil, sand, or mud, and thus are amplified. These amplified seismic waves can greatly change the shape of the land that buildings rest on. *Subsidence* can occur, which means that an area sinks because of settling. Mudflats are also subject to *liquefaction*, which means that the vibration of wet mud or sand results in a "packing down" of the sand grains, forcing the water between the grains to move out. Since water is lighter than sand, the water moves upwards; thus a stretch of mudflats can become a shallow marsh during a severe earthquake. The foundations of buildings that are built on top of mudflats may then be completely submerged.

Bodies of water are also affected by seismic shaking. When an earthquake occurs in or very near an ocean, the shock can create a rapidly moving, powerful seismic sea wave called a *tsunami* (commonly, but incorrectly, referred to as a "tidal wave"). Tsunamis can also enter harbors and cause damage inland from an ocean shore. Lakes and reservoirs are subject to a phenomenon know as *seiche* (pronounced "saysh"), which is similar to the effect achieved by jostling a cup of coffee but on a much larger scale. People who live near dams should be aware that the combined effects of an earthquake's jolt, seiche, and torsion may prove too much for the dam's structural integrity.

Architects and earthquake engineers have learned that a building designed for an earthquake-prone area should have its weight and bulk concentrated near its base. San Francisco's Transamerica Pyramid is an example of this type of building, and it stands little chance of toppling over in an earthquake. Also, the Transamerica Pyramid sits on a foundation that can move independently of the building. Like the weight on a seismograph, the building is expected to remain relatively stationary while its foundation shifts during an earthquake. In contrast, the Seattle Space Needle has a fragile base, and its bulk and weight are concentrated at the top of the structure (see Figure 1 on page 66). Many water storage tanks are elevated on slender legs in this same inadvisable manner. Of more concern in urban areas are the plate glass windows and stone facades of many skyscrapers. These glass and stone plates may shear loose and fall to the ground in an earthquake, creating a significant safety hazard.

Experiences in neighboring earthquake-prone nations have taught us that the best building materials for high-rises are reinforced concrete and steel. Steel is strong and supple, and not likely to break under the stress of seismic waves. All concrete structures must be reinforced with steel rods to prevent the "stack-of-pancakes" result seen in the demise of many unreinforced concrete buildings in Latin America.

Strict building codes in earthquake-prone states also apply to the construction of homes. In California, for example, every new home must have a continuous concrete foundation, 18 inches deep and 12 inches wide, around its perimeter. During an earthquake, the entire foundation might tilt but would not subject the structure to the twisting forces that can destroy a home built with only a corner-support foundation. The best and safest design for a house in an earthquake-prone area is wood-frame construction instead of brick or concrete block. The wood-frame house may twist and flex, but it is likely to remain in one piece. Overall, much can be done to design and construct buildings that will be safe during an earthquake.

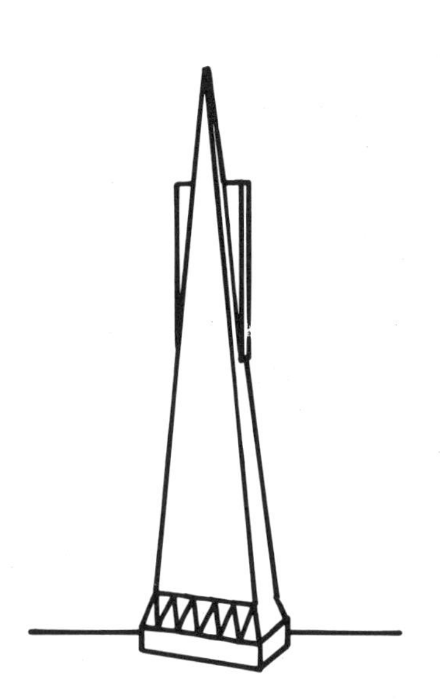

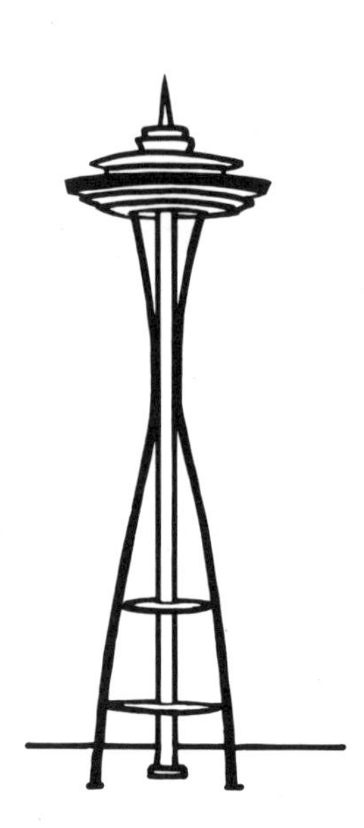

Figure 1. The Transamerica Pyramid in San Francisco, left, has an earthquake-resistant design, while the Seattle Space Needle, right, would be much more prone to severe damage in a strong earthquake.

Lesson Procedure

Step 1:
Review the four types of seismic wave motion (see Lesson 1) with your class. Announce that a simulated earthquake is expected in the classroom today and that students will design a community that will have to undergo the jolts of this earthquake.

Step 2:
Show the prepared sandbox to your students and explain that the yardsticks represent an earthquake fault (see Figure 2). Caution students against jolting the yardsticks accidentally.

Ask students to build a small model community in the sandbox using the materials you've collected. Ask them to be sure to include several types of landforms with different underlying geologic structures. These should include:

- tightly packed areas to represent bedrock
- loosely packed areas to represent landfill or mudflats
- hillsides with slopes of varying steepness

After students have created the various landforms in the sandbox, they should construct their model community. Structures should include:

- tall buildings that are top-heavy
- tall buildings with most of their weight toward the base
- buildings along the buried faultline
- roads that follow hillside slopes and cross the faultline
- power or telephone lines (using straws and string) that extend to all areas and cross the faultline

Step 3:
When everything has been built, discuss your students' expectations of what will happen to the various landforms and structures in the upcoming earthquake. You might have students write down their ideas, or keep a list on the board. Make sure that each student has made a guess about the projected fate of at least one feature in the model community. You might ask for a show of hands indicating agreement after each hypothesis to encourage involvement and commitment.

Step 4:
Simulate an earthquake by jerking both yardsticks sharply in opposite directions with one good, strong jolt (see Figure 3). You might do this with an assistant; or have two students create the earthquake. Caution students not to continue shaking the box after the earthquake.

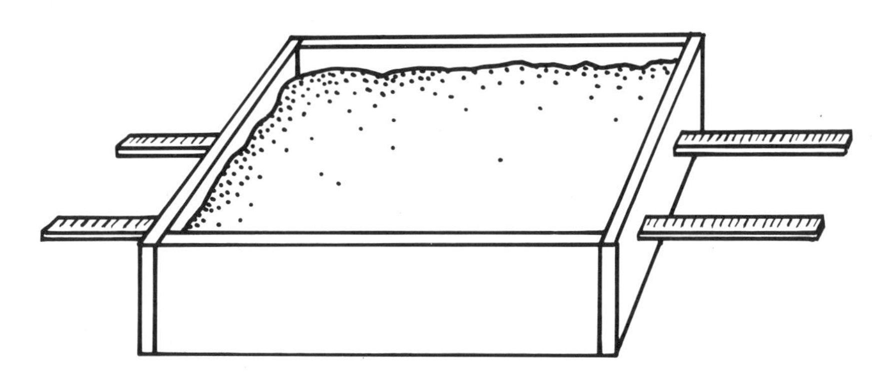

Figure 2. The prepared sandbox for the earthquake demonstration.

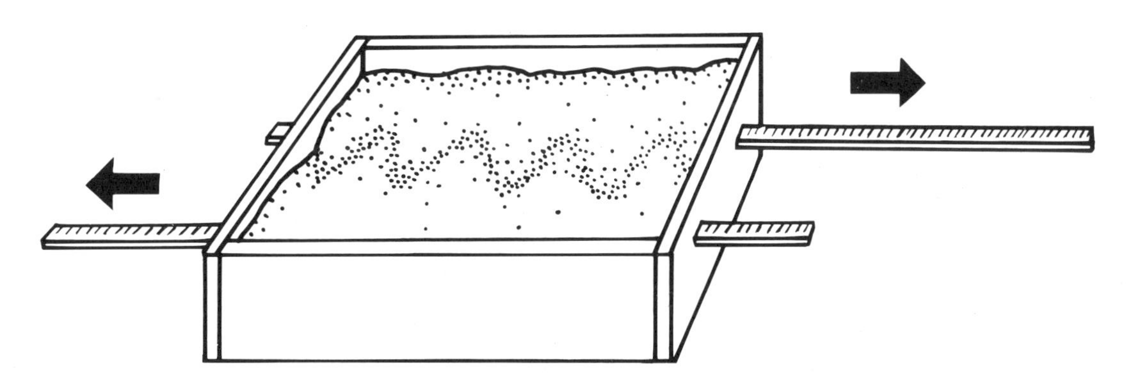

Figure 3. A sharp jolt is applied to the sandbox by quickly pulling the yardsticks in opposite directions.

Step 5:
Ask students to evaluate their hypotheses in light of the results of the earthquake. Did the landforms and structures behave the way students expected them to?

Have students make general statements about the effects of seismic movement on particular areas or structures. Examples might include, "taller buildings topple over," "power lines built over the fault fell down," "hard-packed areas changed shape less than loose-packed areas," etc.

Ask students to apply their generalizations to several landmarks in your area. Go over the information on the transparency, "Landforms and Architecture," or distribute copies of these pages as student handouts.

Step 6:
This part of the lesson is a simple demonstration of liquefaction during an earthquake. Discuss with your students examples of areas built on landfill or mudflats, showing pictures if you can find some in resource books. Then show students the bucket filled with sand and water. Explain that the water fills up all the empty spaces between sand grains at the bottom of the bucket. Carefuly place a block on top of the sand, which represents a building. Ask students for hypotheses about the behavior of the bucket's contents and the fate of the block when it undergoes a simulated earthquake.

Step 7:
Gently vibrate the bucket by drumming on its sides with your knuckles. Very little vibration is necessary to cause the sand level to drop and the water to cover the surface. The block will be swamped, but not submerged (see Figure 4). Discuss the results of this demonstration with your students, and evaluate their hypotheses in terms of the results.

Now that the bucket has a layer of water on top, invite students to name nearby natural or manmade bodies of water (lakes, reservoirs, swimming pools, etc.). Then kick the bucket. Ask students for descriptions of the result. This is an example of "seiche," which could affect dams and dwellings near bodies of water in an actual earthquake.

Figure 4. Demonstrating liquefaction by vibrating the sides of a bucket filled with sand and water.

Enrichment Activities

- Assign your students the task of finding out about the foundations and construction of their homes. Students are likely to find that their homes are well-constructed, given the likelihood of earthquakes in your area, which should be reassuring.
- Find photos of earthquake-damaged buildings in earth science texts or resource books. Ask students for ideas about what could be done to similar structures to make them more earthquake-resistant.
- Students can do research about how your local building code addresses geologic hazards and earthquakes. City Hall, your local government buildings, or the library are good places to start.

Landforms and Architecture

Effects of an earthquake on landforms

Rocks and bedrock

- The denser and stronger the rock, the more shock is absorbed.
  — Waves move faster through denser rock.
  — Faster waves cause less motion.
- The looser and weaker the rock, the less shock is absorbed.
  — Waves move slowly through looser rock.
  — Slower waves cause more motion.

Unconsolidated rock and soil (landfill, alluvium)

- Large, slow waves make the surface unlevel.
- Landslides may occur on hillsides.

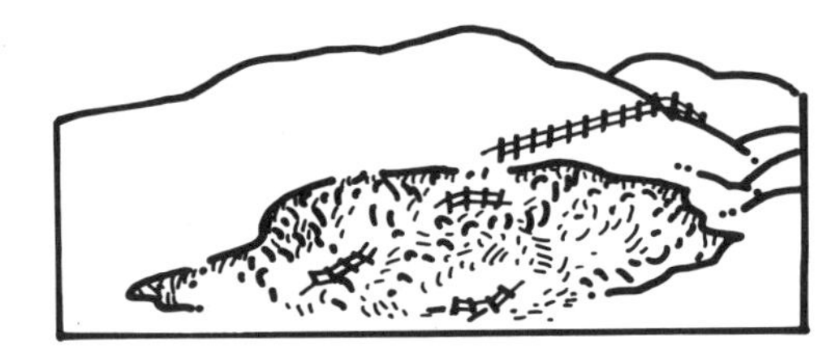

Mudflats

- Have the greatest reaction to earthquake shaking.
- Subject to liquefaction (sand grains settle, forcing water to the top).

Bodies of water

- Seismic sea waves are called *tsunamis*.
  — Ocean water usually recedes from shore, then comes back with great force and height.
- Sloshing of water in lakes, ponds, and reservoirs is called *seiche*.
  — Dams may break, and homes may be flooded.

Effects of an earthquake on buildings

Depends on:

- Distance from an active fault
  — The closer the fault, the greater the damage.
- Soundness of the underlying geologic structure and the building.

Architecture

- Best design concentrates weight and bulk at base, such as a pyramid shape.
- Worst design concentrates weight and bulk at top, such as a water tank or the Seattle Space Needle.

Construction Materials

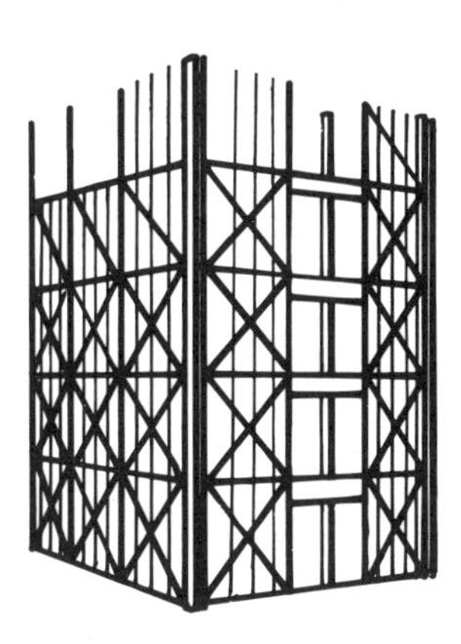

- High-rise buildings should be made of reinforced concrete and steel frames.
  — Steel is flexible and strong.
  — Reinforcing rods within concrete hold the building together even if the concrete breaks up.

- Houses should be made of wood-frame construction.
  — Wood is flexible and strong enough for a house-sized building.
  — The frame may flex, but the walls are likely to remain standing.
  — Continuous concrete foundations 18 inches deep and 12 inches wide are earthquake-resistant. This type of foundation may tilt, but the house probably won't fall down.
  — Less damage will result with a continuous foundation than if the house's four corners were supported separately.

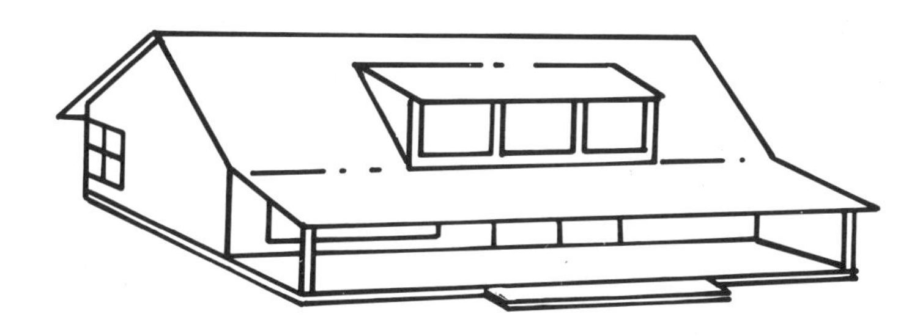

Lesson 4

Keep an Eye on the Tide

Investigating base materials and predicted earthquake intensities

Group Size
Individual students and groups of 10

Time Required
1 class period

Materials
- Generalized geologic maps, enough for three classroom stations. For the San Francisco Bay Region, order U.S. Geological Survey Map MF-709 (contains 3 maps and a cover report) for $4.50 plus $1.50 handling from the U.S. Geological Survey, 555 Battery Street, San Francisco, CA 94111. For other areas, write to: U.S. Geological Survey, Public Information Office, 1028 General Services Administration Building, 19th and F Streets NW, Washington, D.C. 20244. Other sources for geologic maps include your library, local university libraries, city and county offices, and state geologic agencies.
- copy for each student of Handout 1, "Earthquake Risks," on page 76.
- transparency of Handout 1

Key Terms
base material
intensity
prediction
forecast

Instructional Goals
- To help students acquire knowledge about local geologic hazards.
- To enable students to hypothesize about the possible effects of earthquakes on sites that are of interest to them.
- To encourage students to develop family preparedness plans.

Student Objectives
Students will:
- Collect information about base materials (underlying geologic structures) from geologic maps, as well as the predicted maximum intensity of seismic shocks at two building sites.
- Apply their knowledge about the effects of earthquakes on structures and landforms.
- Make generalizations about the likely behavior of base materials at each building site chosen for investigation.

Prerequisite Knowledge
Students should understand the effects of seismic waves on various types of landforms and buildings (from Lesson 3). In addition, students should have a rudimentary knowledge of how to read geologic maps. If these maps are new to your students, spend a few minutes at the beginning of the class period reviewing how to interpret topographic contours and the symbols that indicate faults, buildings, roads, etc.

Advance Preparation Time
About 30 minutes

- Review the "Background Information" and "Lesson Procedure."
- Gather materials.
- Set up one geologic map in each of the three stations. Stations may be desktops or bulletin boards around your classroom.
- Make copies of the student handouts and a transparency of the handout.

Teacher Tips

- Find your school's location on each of the three geologic maps before class and mark it in red ink. This will make it much easier for you to find while demonstrating how to use the maps, and it will be a handy reference for students during the activity.
- Warn students not to mark the maps. The activity can be easily accomplished without writing on the maps.

Background Information

When surfers consult the weather forecast the night before a surfing safari, they are too sophisticated to expect 100 percent accuracy on the part of the meteorologist. The fickleness of weather limits our ability to predict in advance what the weather is going to be like very far in the future, and that is why TV meterologists make forecasts instead of predictions. A forecast is a theory based on data that has been painstakingly gathered and compared with past experience, while a prediction is a statement about the outcome of circumstances before they occur.

Most seismologists are reluctant to predict seismic occurrences and prefer to leave "predictions" to the psychics. Seismologists can, however, forecast the strength and likelihood of seismic activity based on careful observation and comparison of data. In this lesson, students use the same information that seismologists, city planners, public officials, engineering geologists, and earth science consultants use when determining where to locate housing developments, schools, hospitals, parks, etc. Since my school is located near the famous San Andreas Fault, I have geared this lesson toward the seismically active San Francisco Bay Area. A great deal of geologic and earthquake information is available for this region, making it easy to study. You might tailor this lesson toward your own location—U.S. Geological Survey maps are available for the entire country—or you might use the San Francisco Bay Area maps as an exercise for your students. In either case, the lesson is designed to help your students reach conclusions about the possible effects of an earthquake on various buildings.

In the course of this activity, your students may discover that the geologic maps do not contain all good news. Many homes, as well as commercial and public buildings, are likely to experience a lot of shaking during an earthquake, and the land under them may change shape. In many instances, poor zoning decisions have allowed developers to build near active faults, sometimes even directly on top of faults. For this reason, the knowledge your students gained in Lesson 3 about earthquake effects on structures and landforms may provide reassurance and power. The purpose of this activity is to empower your students with knowledge, and to motivate them to discuss the information with their families in order to make emergency procedure preparations.

Lesson Procedure

Step 1:
Review with your students the information learned in Lesson 3 about the effects of an earthquake on landforms and structures. It is particularly important that your students understand the relationship between the base material, or underlying geologic structure, and suitable architectural designs.

Step 2:
Ask students for further observations or thoughts about their community, its topography and geology, and its architecture.

Brainstorm with your class, listing locations of buildings that would be of concern to your students in the event of a major earthquake. Record their responses on the chalkboard. Responses typically include their school, homes, parents' workplaces, hospitals, and civil defense centers.

Discuss with your class how seismologists are able to make reasonable forecasts of earthquake shock intensity by using mathematical formulas based on historical records, geologic composition of different sites, and distance from an active fault.

Step 3:
Introduce the three maps and three classroom stations. The maps contain information collected for the San Francisco Bay Area, or your local area, and will be used by students to make forecasts and do subsequent planning activities. If you use the San Francisco maps, be sure to tell your students that the information contained on the U.S. Geological Survey maps is based on data collected after the San Francisco earthquake of April 18, 1906. The "Predicted Intensities" are based on reports of damage, as well as the relationship of the amounts of damage to the geologic character of the ground and the distance from the fault zone.

Distribute the student handouts, and place the transparency on an overhead projector. As an example, select one site from your students' brainstorm list, such as your school. Demonstrate how to locate the school on the map and how to read important information on the map, such as fault lines, geologic formations, etc. To complete the handouts, your students will need to identify the generalized geologic composition under a site (what sort of rock the building is built on); the "Predicted Maximum Intensity" at the site (how big a jolt is expected in that area), and the "Predicted Maximum Intensity" at the nearest "Specific Site" (how big a jolt is expected at the nearest seismic station, based on past experience). If all this information is not available on your local maps, adapt the student handout to include as much relevant information as you can, such as distance from a fault, underlying geologic formation, etc. Record your findings on the transparency to show your students the correct procedure for completing the handout.

Step 4:
Direct each student to choose two sites from the brainstorm list to investigate in this activity, just as you did on the transparency.

Divide the class into three groups and instruct students to rotate through the three map stations, collecting information from each map about the locations they have selected. Provide assistance where necessary and encourage peer assistance.

Step 5:
Compare and discuss individual results. Note sites of maximum and minimum projected earthquake severity. You might point out the types of base materials identified on your geologic maps, such as stable bedrock formations or unstable formations.

Step 6:
Ask each student to generalize about the effects of a major earthquake on one or both sites, including hypotheses about the possible behavior of the base materials when subjected to the earthquake shock likely at that site. Orally or in writing, students should be able to make statements like, "The hospital is close to the Hayward Fault, but it's built on solid rock. The predicted shock at that site is less than at places made of looser rock."

Enrichment Activities
Students can:

- Collect further data for each of the sites chosen in the "Earthquake Risks" activity. Information about the construction of buildings, foundation materials, and actual base materials should be available from local agencies. For example, the geologic map may indicate bedrock, but students might investigate whether fill has been used to level the lot.
- Discuss earthquake preparedness plans with their families, then share these plans orally in class or in a written report.
- Present the information, either orally or in writing, that they have gathered from geologic maps as evidence of a need for expanded public policy and preparedness.
- Write letters to public officials or local newspapers regarding the locations of important public buildings in relation to seismic hazards.

Earthquake Risks

Choose two sites and answer the questions below using the geological survey maps.

Name of Site 1: ____________________

1. Name the nearest active fault to Site 1: ____________________
2. Distance from this fault to Site 1 in km: ____________________
3. Find the Specific Site (seismic station) nearest to Site 1 and record its location: ____________________
4. Find the Predicted Maximum Intensity at this Specific Site and circle it.

   \+ 4 3 2 1 0 –
5. Determine the Generalized Geologic Composition of Site 1:

   ____________________ which stands for ____________________.
   (symbol) (type of material)
6. Find the Predicted Maximum Intensity at Site 1 and circle it.

   \+ A B C D E –

Name of Site 2: ____________________

1. Name the nearest active fault to Site 2: ____________________
2. Distance from this fault to Site 2 in km: ____________________
3. Find the Specific Site (seismic station) nearest to Site 2 and record its location: ____________________
4. Find the Predicted Maximum Intensity at this specific site and circle it.

   \+ 4 3 2 1 0 –
5. Determine the Generalized Geologic Composition of Site 2:

   ____________________ which stands for ____________________.
   (symbol) (type of material)
6. Find the Predicted Maximum Intensity at Site 2 and circle it.

   \+ A B C D E –

Predators and Pollinators
A Study of Natural Selection

Jean Lyford
Orinda Intermediate School
Orinda, California

I have taught science and English to seventh and eighth grade students in the Orinda Union School District since 1972. As part of my biology curriculum I include units on ecology and genetics, which lead to a study of natural selection, the topic of my unit in this book.

To me, teaching is a life-enhancing profession that stimulates growth. Before moving to California, I taught four-year-olds in a private school in New York City, and then switched to teaching seventh through ninth grade at a public junior high school in Brooklyn. My first teaching assignment with four-year-olds taught me to admit vulnerability in order to be open to potential growth in myself and my students. My young students' openness, originality, curiosity, and joy in discovering themselves helped me recognize my own beginnings, and my potential for continued growth.

At the end of my first semester in the Brooklyn junior high, however, I was discouraged and wondered if some of my eighth grade students would ever learn to read. One incident in particular helped me discover that my four-year-old students had taught me a valuable lesson: we are all really four years old at heart—we just try to hide it. Ramon, one of my non-reading students, was caught trying to set fire to auditorium curtains that were clearly marked with the word asbestos. *Ramon was brought to me, and I pointed out that he would not have had to go through this if he had been able to read. By the end of the year Ramon was reading. Orinda, California, may be far away in both distance and culture from Brooklyn, but kids are kids—still vulnerable, curious, and open to growth.*

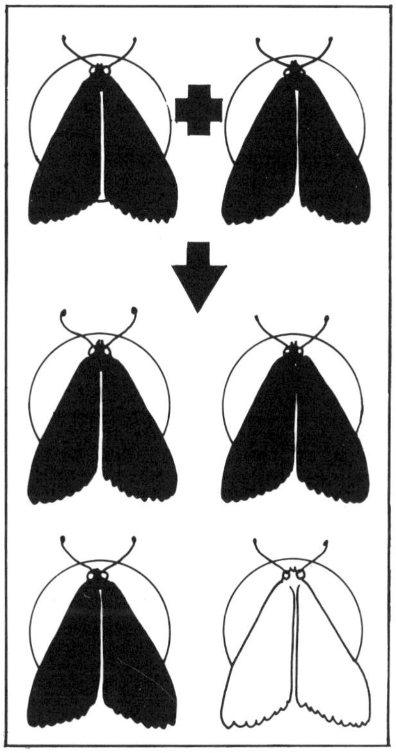

Lessons:
1. "Like Two Peas in a Pod?"
2. "Black or Clear: Action or Inaction"
3. "It Takes Two to Tangle"
4. "T-Shirt Mania"
5. "The Birds and the Bees"

Overview

THE PEPPERED MOTH, native to the area surrounding Manchester, England, provides a fascinating study in natural selection and the predator/prey relationship. The peppered moth is active only at night and rests on the bark of the local sycamore trees during the day. In the mid-1800s these spotted moths varied in color from almost white to dark grey or brownish-black. The moths that were lighter blended well with the variegated grey-white sycamore bark, camouflaging them from predatory birds. As a result, most of the moths were light in color. When moth collectors found one that was almost black they treasured it as a rare prize. Less than 10 percent of the moths were dark, but darker-colored moths continued to exist. And so, for generation after generation of moths, the relationship between the moth, the tree, and the predator remained in balance.

Then something changed the environment to upset this natural balance. The change was caused when Manchester became industrialized, polluting the air with soot from coal-burning factories and homes. The sycamore bark no longer was a blend of light greys and pale tawny colors; instead the trees had a dark sooty color. Thus, the light-colored peppered moths were very obvious during the day because they stood out against the sooty bark (see Figure 1 on page 80).

A strange phenomenon then happened during the next 75 years. People who collected moths found it was no longer a rarity to find a dark peppered moth. In fact, the percentages had been reversed. Now less than 10 percent of the moths were white; most moths were dark. Despite the change, however, some white moths still contributed their genes to the population's gene pool.

The story is not over, however. Within the next 50 years, the city of Manchester succeeded in cleaning up its pollution. Black smoke no longer bellowed out of factories, and the sycamore tree bark returned to its original light color. What happened to the dark moths? They were no longer well-adapted to their environment because they stood out on the light, soot-free bark. Today, after two color changes, the peppered moths are back to their original color.

The story of the peppered moth is an example of natural selection through the predator/prey relationship in a changing environment. Normally such changes take place over many thousands of years. One reason these changes occurred so quickly in the peppered moth is because of the moth's short life cycle. Another reason is human intervention through pollution and subsequent clean-up of the environment.

The important concepts to be learned from this example are twofold. First, variation in the gene pool of a population contributes to the survival of the fittest. Although light-colored moths originally had an advantage, individual moths always ranged in color from pale to dark. Even though most were light, a few dark moths survived long enough to reproduce. If all the moths had been pale with no dark survivors, the species would have been unable to survive when the environment changed. The reverse became true when the tree bark lightened. The genetic variation in the gene pool safeguarded the potential for adaptaion to a changing environment.

The second important concept illustrated by the peppered moth tale is that a dominant trait can become less common if a recessive trait is better adapted for survival in a particular environment. Further, the predator/prey relationship is an important factor in selecting the trait that will survive. Even though dark-colored moths were dominant over light-colored moths, fewer dark moths survived long enough to reproduce because the predator birds were better able to find the dark moths on pale bark. Fortunately for the peppered moth species, some dark moths did escape the predators. The dominant trait still existed in the gene pool and was able to reappear when conditions were right.

Several textbooks mention the peppered moth story (see "Extensions and Sources"). If you use these books with your students, point out that some of the books tell only the first half of the story. Scientists have recently discovered that, as a result of a reduction in

Figure 1. Light and dark peppered moths resting on a sycamore tree with dark-colored bark.

pollution, the moth has returned to its original color. It is a tale with a happy ending.

While the peppered moth story is presented here to give you an introduction to the unit, I find it especially useful to read it to my class *after* completing the unit. You might have your students write a composition in which they thoroughly explain the reasons behind the two changes in moth color. This composition will provide you with a good evaluation of your students' understanding of the relationship between genetics and natural selection.

This unit is designed for use after your students have already learned the principles of genetics and are familiar with the theory of evolution. Its purpose is to synthesize the concepts learned in genetics with the process of natural selection. Students often misinterpret adaptation over a long period of time as meaning an organism "gets used to an environment slowly." If students understand that the genetic code carries traits that are passed on to the next generation, they can begin to appreciate the outside forces that contribute to the survival of certain individuals, populations, and species.

Of all the life sciences, genetics has produced some of the most controversial social issues of our time. The implications of recombining genes, "test tube" (*in vitro*) babies, and genetic counseling for birth defects will be researched, discussed, and argued for many years to come. Since your students will be affected by these issues, it is necessary that they have the background information to understand the far-reaching implications of how genetic engineering may shape our future.

This unit is intended to provide students with an understanding of the impact genetics has on our everyday lives. Variation is not only important with moths, but for human beings, too. What does the loss of a species due to human tinkering do to the biosphere as a whole? Should people be able to choose the sex of their children? If so, how will this affect the probabilities for future social structures or even survival of our species? If we change our environment, will we be able to adapt to that change? Perhaps we can adapt, or at least some of us can, to a minor change. But is our human gene pool strong and varied enough to withstand a catastrophic change? The dinosaurs didn't make it.

Key Concepts

- ☐ Genetics and natural selection are interrelated. Students should understand these important terms:
  - variation
  - dominant and recessive traits
  - predator/prey relationship
  - adaptation
  - phenotype/genotype
  - population
  - gene pool
  - mutualism
  - natural selection
- ☐ The study of genetics can produce a greater understanding of how organisms adapt to environmental changes, leading to many implications for our future.

Skills Used in the Lessons

- gathering and recording data
- classifying
- measuring and graphing
- comparing quantities
- matching and grouping
- making inferences
- using space-time relationships
- controlling and manipulating variables
- recognizing and predicting patterns
- synthesizing and stating laws

Extensions and Sources

In order to verify the information in this unit and complete the story of the peppered moth, I talked with my friend, Dr. William Libby, Professor of Genetics and Forestry at the University of California, Berkeley. He brought me up-to-date on the story of the peppered moth. His eyes sparked with glee as he told me of the moth's return to its original color. It's good to know that some mistakes are reversible.

Students interested in pursuing careers focusing on genetics can look into medicine, genetic engineering, crop management, pharmacology, biochemistry, genetic counseling, forestry, etc. These exciting professions are not only interesting to individuals, but also benefit the rest of mankind.

This unit could easily be extended to include mutations as contributors to variation, genetic diseases, sex-linked characteristics such as color-blindness, mixed dominance, multiple gene inheritance, differences between heredity and environment, social impact of genetic engineering, and the impact of pollution. Information on these topics can be found in the books listed below.

Berkow, Robert. *The Merck Manual.* Rahway, NJ: Merck, Sharp, and Dohme Research Laboratories, 1982.

Bierer, L., and Fisher Lien, L. *Life Science.* Washington, D.C.: Heath and Co., 1984.

Brandwein, P. *Life: Its Forms and Changes.* New York: Harcourt Brace Jovanovich, Inc., 1972.

Brandwein, P., et. al. *Life: A Biological Science.* New York: Harcourt Brace Jovanovich, Inc., 1975.

Capra, J., Corley, K., et. al. *Genes and Surroundings.* Dubuque, IA: Kendall/Hunt Publishing Co., 1983.

Haffner, R. *Genetics—The Thread of Life. New Biology for Young Students.* Middletown, CT: Wesleyan University, 1964.

Minnesota Environmental Sciences Foundation, Inc. *Genetic Variation.* Washington, D.C.: National Wildlife Federation, 1972.

Ramesy, L., Gabriel, L., et. al., *Life Science and Health.* New York: Holt, Rinehart and Winston, 1978.

Glossary

This glossary contains all the key terms contained in the lessons, as well as some terms that might be helpful as background information, particularly if you have students who may have heard these words and want them clarified.

Adaptation: The ability of an organism, through structure or behavior, to be successful in a given environment. "Successful" means the ability to survive and reproduce.

Alleles: Genes that occupy the same location on chromosomes. Alleles are paired to produce a genetic trait.

Chromosomes: Rod-like structures in the nucleus of cells that carry genes.

Dominant trait: A characteristic that is exhibited regardless of whether one gene (allele) or both genes carry the trait.

Gene: The basic unit of heredity that is made up of molecules of DNA.

Gene pool: All the genes in a population available for crossing (breeding). The gene pool includes all the inheritable traits of a population.

Genotype: The actual combination of genes in an organism.

Heterozygote: An organism having two different alleles for a given trait.

Homozygote: An organism having identical alleles for a given trait.

Mutualism: The relationship between two different species in the same environment that helps one another survive; i.e., the bee gets food from the blossom and at the same time pollinates the plant.

Natural selection: Organisms with traits enabling them to adapt to their environment are more likely to survive and reproduce than those organisms that are poorly adapted.

Phenotype: The appearance of inherited traits in an organism; traits that can be observed.

Population: A group of individuals of the same species in the same environment.

Predator/prey relationship: The interaction of two species in the same environment where one feeds on the other.

Probability: The mathematical prediction of odds in reordering possible numerical combinations.

Recessive trait: A characteristic that is not exhibited unless both genes (alleles) carry that trait.

Species: The most specific level of taxonomy (classification of living things). Includes those individuals that can mate and produce fertile offspring.

Survival of the fittest: Darwin's theory of natural selection. States that all organisms, even of the same species, show variations of traits that are passed to the offspring. Most organisms produce more young than can survive; thus, every organism must constantly struggle to survive through adaptation. Those that are most fit (adapted) will survive to reproduce.

Variation: An organism's divergence from the typical qualities or genetic characteristics of its species.

Lesson 1

Like Two Peas in a Pod?

A study of variation in a sample population of black-eyed peas

Group size
Self-chosen groups of two

Time Required
1 class period

Materials
- enough black-eyed peas to fill a styrofoam cup for each group
- one styrofoam cup for each group, filled to the rim with peas
- one plastic or metal tray (TV trays work well, but they must have raised edges) for each group
- one metric ruler for each group
- one sheet of graph paper with 1/4" or 1/2" squares for each student
- copy of Handout 1, "Peas in a Pod" on page 88 for each student

Key Terms
- population: all the black-eyed peas in the room
- variation: differences between the individual peas
- species: all black-eyed peas that could cross-pollinate and produce fertile offspring

Instructional Goal
- To observe and appreciate the diversity of individuals within a population.

Student Objectives
Students will:
- Sort peas according to observed differences, and classify those differences.
- Record, measure, and graph the results of these observations.

Prerequisite Knowledge
Students should have had a unit on basic genetics prior to the lessons in this unit. Students should understand concepts such as inheritance, variations, and genes. You might spend a few minutes at the beginning of the lesson reviewing what your students already know, or explaining any unfamiliar terms.

Advance Preparation Time
About 45 minutes
- Review the "Background Information" and "Lesson Procedure."
- Gather materials.
- Make copies of the student handouts.

Teacher Tips
- Make sure the TV trays or plastic trays used in the activity have raised edges, since you don't want any stray peas around your classroom.
- Mark a line around the inside rim (about 1/4" or 1/2" from the top) of each styrofoam cup, which will indicate the top level of the peas. This should help control the number of "lost" peas. Also, if you are teaching this lesson to more than one class, you can simply fill the cup to this point for each following class.
- Be alert for creative pea shooters as the class period progresses.

Background Information

Most junior-high age students go to great lengths to separate themselves from the adult world, but they are careful to conform to their peer group. Even when their attire or behavior is outrageous, they maintain a certain conformity. The "in" groups and "out" groups may be different in appearance from one another, but there is little room for variation within the group. All the "punks" have a certain style, as do the "populars." As a result, one of the most difficult concepts for students of this age to understand is the importance of *variation* (diversity) for the survival of any population. The purpose of this lesson is to have your students observe, qualitatively and quantitatively, the diversity of black-eyed peas. This is the starting point for their understanding of genetic variation.

Pea plants have long been favored by geneticists for experimental work, primarily because peas are self-pollinating, making it easier to produce a pure strain. When we say two people are "like two peas in a pod," it means they are very much alike. But just how similar are peas in a pod? We expect brothers and sisters to be somewhat alike. In their genetic make-up peas are more like siblings born to first cousins. Even though peas of a pure strain are more closely related than human siblings, it is amazing how different each pea is from the other.

When beginning the activity in this lesson, your students will probably observe the more obvious differences in a sample population of black-eyed peas, such as size and color. But since the activity requires students to find three classifications of variation in the population, they'll have to come up with more than just those two. Some students will find unique differences, such as the size of the light spot in the black eye, or a ratio of the size of the eye to the size of the pea, etc. Student observations will be quite different from group to group. Some students will be content with the obvious differences, but others will come up with observations you never thought of yourself. You'll find it delightful to share in their creativity.

Lesson Procedure

Step 1:

Distribute the student handouts. Have students pair up in groups of two, and ask one member from each group to get the supplies you have set out (these supplies are also listed on the student handout). Point out that the cup of peas is the group's sample population and that they are looking for variations within that sample.

Step 2:

Students should read through the handout and follow the directions. Monitor students as they decide which characteristics of peas they will select to write in their charts. Those who can't find characteristics other than color, size, or texture might need a little prodding with a question such as, "What other differences do you see?" Make sure at least one variable can be measured with the metric ruler.

Taking one variable (difference) at a time, students should sort their peas on TV trays, then record the number of peas showing that characteristic on their charts. One variable might be, for example, "Size: larger than 5 mm." (See the sample chart on page 86.)

Step 3:

Students will take the results for one (or each) of the variables and plot the number of peas on their graph paper. A bar graph works well for this. Some students will need help when they reach this point.

You might draw a model graph on the chalkboard as an example.

Since some students will complete the sorting, classifying, and plotting before others, have those students figure out the percentages and record them on their graphs.

Step 4:
When your class has completed the assignment, have each group share their results. Write the results on the chalkboard, then discuss the similarities and differences between the findings of each group. Elicit from students not only what those differences are, but what they might mean. Why are some peas wrinkled and small? Why are there differences between the samples for the same variation? (One sample may not be representative of the population as a whole.) With gentle prodding on your part, your students can come up with many inferences.

It is my custom to call on every student during class periods that involve discussion. As a result, I get a good idea of which students need reinforcement and which students understand the concepts. I still call on students who volunteer, but I also make sure to call on those who don't.

Enrichment Activities

- ☐ To reinforce the concepts of the lesson, as well as to give students the opportunity to do the activity on their own, have them each take home a new piece of graph paper and another copy of the handout. They should use the same instructions on the handout to examine a population of grain or dried beans from their kitchen cupboards. Possible populations could include coffee beans, rice, barley, or dried beans of any kind (navy, limas, peas, lentils, or kidney beans). Most households have at least one of these items. For this assignment, you might give extra

Variation	Size - larger than 5 mm	Color - more dark than light	Size of eye - up to 2 mm
Number of peas with that variation	113	37	84
Total number of peas in your population	186	186	186
Percentage of peas in population with that variation	61 %	20 %	45 %

Sample student chart for Handout 1.

credit for figuring out the percentages, particularly if your students have difficulty with this. Your primary goal, however, is to have students appreciate the diversity of a population and record their observations in an organized way, not to spend a lot of time learning how to do percentages.

- As an extension of the activity, bring about a dozen intact, fresh pea pods to class. Have each group examine a pod to find reasons for the variations in pea size. When students open the pods they will notice that the peas in the center are larger. Why? Are the peas at the ends constricted? This could lead to a discussion about the difference between heredity and environment, perhaps even extending into a science fair topic or a library research project. How could students test for the importance of environmental versus genetic factors? (For example, if peas from the same pod were dried and planted under the same conditions, would the small peas produce small offspring?)

Peas in a Pod

Variation in a Population of Black-eyed Peas

Get the following materials from your teacher for your group:

- a cup of black-eyed peas
- a metric ruler
- a tray
- two sheets of graph paper

Although black-eyed peas look very much alike at first glance, they are actually very different. Think of your cup of peas as a sample of the total *population* of black-eyed peas in the room. Carefully pour the peas onto the tray so you can look at all the peas at once. With your partner, look for ways the peas are alike and ways they are different. Then choose three differences, including at least one that can be measured with your ruler. These will be the three *variations* you record in the chart below. After you've finished the chart, use the graph paper to make a bar graph of your results for one of the variations.

Variation			
Number of peas with that variation			
Total number of peas in your population			
Percentage of peas in population with that variation*			

* To get the percentage, divide the number of peas in your sample that show the variation by the total number of peas in your population, then multiply by 100. For example, if 54 peas out of 132 peas show a particular variation, then the percentage is $54 \div 132 = 0.409 \times 100 = 40.9\%$

Lesson 2

Black or Clear: Action or Inaction

How dominant and recessive traits contribute to variation

Group Size
Individual students

Time Required
1 class period

Materials
- 2 pieces of black or dark-colored construction paper for each student (you can cut an 8-1/2" by 11" sheet of paper in half lengthwise to produce two 4-1/4" by 11" strips)
- 2 pieces of clear plastic strips (similar to transparency film), the same size as the construction paper, for each student
- copy for each student of the two-page handout, "Black or Clear?" on pages 93–94
- transparency of the two-page student handout, with answers filled in (see answers on page 91)

Key Terms
dominant trait
recessive trait
genotype
phenotype

Instructional Goal
To promote student understanding of how dominant traits, recessive traits, genotypes, and phenotypes help produce variation in a population.

Student Objectives
Students will:
- Manipulate strips of paper and plastic to represent the inheritance of dominant and recessive genes.
- Chart phenotypes and genotypes, as well as dominant and recessive traits.
- Write an analysis of how dominant and recessive traits contribute to variation in a population.

Prerequisite Knowledge
Students should be familiar with the basic principles of genetics, as well as the material in Lesson 1.

Advance Preparation Time
About 35 minutes
- Review the "Background Information" and "Lesson Procedure."
- Gather materials.
- Make the transparencies and student handouts.

Teacher Tips
- Some textbooks do not use the terms *genotype* and *phenotype*. If this is the case with the book you used for your unit on genetics, these words will have to be introduced for this lesson rather than reviewed. The genotype is simply the actual paired genes that have been inherited for a particular trait. The phenotype is the inherited characteristic that is evident in the physical appearance or behavior of the individual. If a dominant gene is present in the genotype, it is manifested in the phenotype. The phenotype can only reflect the recessive trait when no dominant genes are present (see the glossary on pages 82–83).
- Although students will work individually in this lesson, I let them share responses so they can learn from one another.

Background Information

Most textbooks define dominant traits as those that are expressed in the appearance or behavior of an organism, and recessive traits as those that are hidden or masked by the dominant trait. New research techniques, such as mapping genes on the chromosomes and recombining genes, have led scientists to a slightly different interpretation of the relationship between dominant and recessive traits. Rather than simply masking the recessive trait, the dominant trait acts—it exhibits itself. The recessive trait, on the other hand, is inactive.

For example, suppose the peeping behavior of chicks represents a dominant characteristic. Then chicks that are unable to peep would represent the recessive characteristic. It's not that the characteristic of being unable to peep is hidden by the gene that carries the peeping trait. Either the chick can peep or it can't peep. This means that if the dominant characteristic appears at all, it acts. In this case the chick peeps. The only case in which a silent chick is produced is when both genes are unable to act—both genes are recessive. The peeping behavior in chicks is an advantage, for it aids them in bonding with the parent birds and in making their various needs known. If this behavior was to become a disadvantage—for example, if bonding became less important than being silent and hiding from predators—then the silent chicks would be better adapted than the peeping ones.

In this lesson students learn about dominant and recessive traits by combining clear plastic strips and colored construction paper strips, which simulates various genotypes and phenotypes. The clear plastic strips represent the inactive recessive gene, while the colored construction paper represents the active dominant gene. If you prefer to use the textbook definitions of dominant and recessive traits, they will not affect this lesson. The distinction between showing and acting, as well as hiding and not acting, may seem like a subtle one; however, it is important for students to understand that scientists change their hypotheses and theories as new evidence comes to light.

Lesson Procedure

Step 1:
Review the definitions of the Key Terms listed on page 89 before your students begin the activity. Discuss with your students the differences between dominant and recessive traits, as well as the difference between the terms phenotype and genotype. Elicit from them how the latter terms are related to the former. For easy reference you might write the definitions of these terms on the chalkboard, or have your students write the definitions on notepaper.

Step 2:
Distribute the two-page student handout and give each student two plastic strips and two pieces of colored construction paper. Go over the instructions on the handout, and have students complete both pages.

Step 3:
After all students have completed the handout, place the transparency on the overhead projector so they can check their work. Answers are shown on page 91.

Answers to Handout 1.

1. 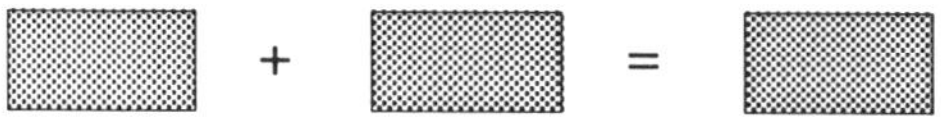 + =

The phenotype is dark; the genotype is DD.

2.  + =

The phentoype is dark; the genotype is Dd.

3. 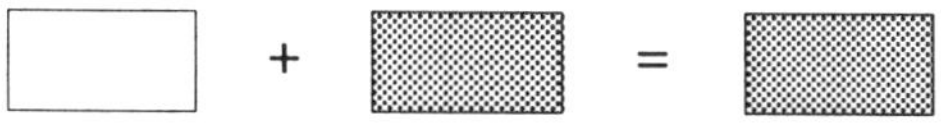 + =

The phenotye is dark; the genotype is dD.

4. 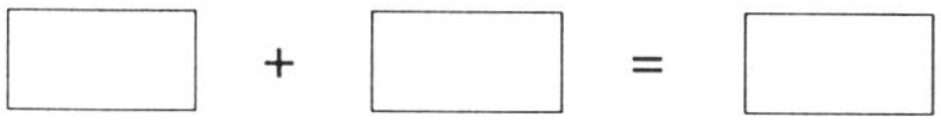 + =

The phenotype is clear; the genotype is dd.

				Phenotype	Genotype
5. Dominant D	+	Dominant D	=	dark	DD
6. Dominant D	+	Recessive d	=	dark	Dd
7. Recessive d	+	Dominant D	=	dark	dD
8. Recessive d	+	Recessive d	=	clear	dd

9. The characteristic that always shows if it is present in the genes is called a dominant trait. It is the phenotype. Even if only one gene for a dominant trait is present, it will be the phenotype. The only way a recessive trait can be a phenotype is if both (or two) genes carry the trait. The actual combination of genes is called the genotype.

10. (Refer to the chart showing two generations of moths.)
What characteristic is dominant? black What characteristic is recessive? white
The mating of the first two moths will always produce what phenotype? black
What genotype? Dd The mating of two moths that each carry a recessive and dominant trait can produce how many genotypes? 3 (Note: Some students may say 4 genotypes, because technically Dd is different than dD. The father's gene is labeled first, and the mother's gene is second, so that the D would be from the father in the first genotype and the d would be from the father in the second genotype.)
The parents of a white moth must have what genotype? dd or dD or Dd What would be the possible genotypes of a parent of a black moth? DD, Dd, dD, or dd

Step 4:
After going over the handout, discuss dominant and recessive traits with regard to the moths. Ask your students why there might be a difference between the genotypes Dd and dD. What is the advantage of agreeing on a system that tells which parent comes first? (By stating the order we are able to tell the genotype and also which allele came from which parent. One application of this is in discovering carriers of inheritable traits and diseases.)

Step 5:
Have students write a paragraph explaining what they've learned (see the bottom of the second page of the handout). This can be done either as homework, or in class if time permits.

Enrichment Activities

- ☐ Using their textbooks or library resources, students can list dominant and recessive traits that are paired. This activity can include making a pedigree of a trait, whether in plants, humans, or other animals.
- ☐ Have students do research on an inheritable trait or disease, including how scientists found the responsible gene and, in the case of a disease, how genetic testing and counseling can prevent the disease.

Black or Clear?

Obtain two pieces of dark-colored construction paper and two pieces of clear plastic from your teacher. The dark paper represents a dominant gene, and the clear plastic represents a recessive gene. In the following exercises, use the words "dark" and "clear" to describe the phenotype, and the letters "D" for dominant and "d" for recessive to describe the genotype.

1. Place one dark-colored strip on top of the other dark-colored strip. Shade in the blank rectangles below to show this combination of genes, then fill in the blanks in the sentence.

   ☐ + ☐ = ☐

   The phenotype is _______; the genotype is _______.

2. Place one dark-colored strip on top of one clear plastic strip. Shade in the rectangles to show the results, then fill in the blanks in the sentence.

   ☐ + ☐ = ☐

   The phenotype is _______; the genotype is _______.

3. Place one clear plastic strip over one dark-colored strip. Shade in the rectangles to show the results, then fill in the blanks in the sentence.

   ☐ + ☐ = ☐

   The phenotype is _______; the genotype is _______.

4. Place one clear plastic strip over the other clear plastic strip. Does any color show? Leave the right-hand rectangle blank, since the dominant trait for the dark color is not present. It does not act in this combination of genes.

   ☐ + ☐ = ☐

   The phenotype is _______; the genotype is _______.

Shade the appropriate rectangles below, and write the letter "D" next to the word "Dominant" and the letter "d" next to the word "Recessive." Decide what each phenotype (dark or clear) and genotype (DD, Dd, dD, or dd) will be, and fill in the blanks.

						Phenotype	**Genotype**
5. Dominant	___	+	Dominant	___	=	________	________
6. Dominant	___	+	Recessive	___	=	________	________
7. Recessive	___	+	Dominant	___	=	________	________
8. Recessive	___	+	Recessive	___	=	________	________

Write the missing words in the blanks.

9. The characteristic that always shows if it is present in genes is called a ________ trait. It is the phenotype. Even if only one gene for a dominant trait is present, it will be the ________. The only way a recessive trait can be a phenotype is if ________ genes carry the trait. The actual combination of genes is called the ________.

10. Study the chart showing three generations of moths. What characteristic is dominant? ________ What characteristic is recessive? ________ The mating of the first two moths will always produce what phenotype? ________ What genotype? ________ The mating of two moths that each carry a recessive and dominant trait can produce how many genotypes? ________ The parents of a white moth must have what genotype? ________ What would be the possible genotypes of a parent of a black moth? ________

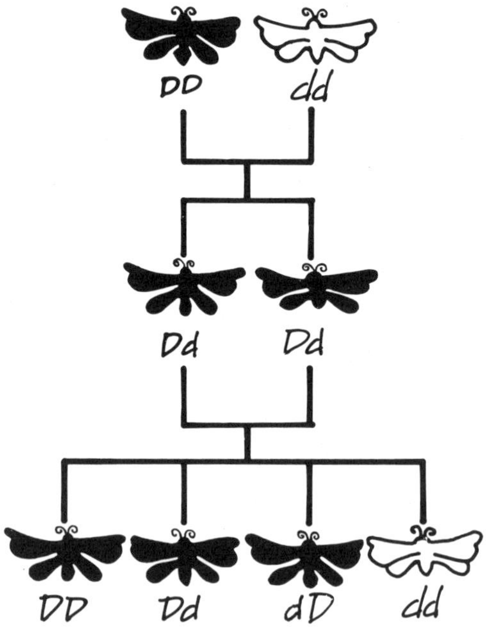

Using what you have learned so far, write a paragraph explaining how the inheritance of dominant and recessive genes helps produce variation in a population.

Lesson 3

It Takes Two to Tangle

Simulating a population with dominant and recessive traits

Group Size
2-4 students

Time Required
1-2 class periods

Materials
- 1 pair of dice for each group; half white and half red. I usually borrow dice from board games at home; however, dice can also be purchased through educational suppliers or discount stores.
- 1 package of small gummed labels (to be cut and pasted on each face of the dice)
- 1 piece of felt, each about 12 inches by 12 inches, for each group (optional). You can usually find inexpensive remnants at a fabric store; neither texture nor color matters.
- copy of Handout 1, "Population Worksheet" for each group
- copy of Handout 2, "Gene Pool Summary Sheet" for each student

Key Terms
gene pool
probability
dominant/recessive trait
genotype
phenotype

Instructional Goal
- To recognize how the genetic probability of dominant and recessive gene inheritance affects variation in a population, and to emphasize that a trait requires one gene from each parent.

Student Objectives
Students will:
- Roll dice to build a simulated population of males and females that combines genes for a given trait. Offspring will be identified according to sex and one dominant/recessive trait.
- Compare their population results with the mathematical probability of genetic inheritance of dominant/recessive traits.

Prerequisite Knowledge
Students should be familiar with the material in Lessons 1 and 2. In addition, students should have a basic understanding of probability.

Advance Preparation Time
About 1-1/2 hours
- Review the "Background Information" and "Lesson Procedure."
- Gather materials.
- Cut the small gummed labels into pieces just slightly smaller than the face of a die. You will need enough pieces to cover the faces of all the dice.
- Paste the labels onto the dice. Use a marking pen to write a "D" on two faces of each die and a "d" on two faces of each die. Then write an "X" on two faces of each white die. On each red die, write an "X" on one face and a "Y" on one face.
- Duplicate the student handouts.
- On your chalkboard (or on butcher paper or a transparency), write the table from the top of the Handout 2, "Gene Pool Summary Sheet."

Teacher Tips

- Divide the students into groups of 2-4, depending on your class size and availability of materials (groups of 2 work best).
- Have each group roll its marked dice on a felt cloth. It is not absolutely necessary to use the felt, but I have found it cuts down on noise.

Background Information

After learning in Lessons 1 and 2 about the importance of variation in a species and a population, your students are ready to use these concepts to form their own simulated population. In Lesson 1 students simply observed a ready-made population of black-eyed peas. In this lesson students will randomly combine genes carrying a trait to produce a population with a finite gene pool.

Your students have probably heard of the "law of averages" with regard to rolling dice or tossing a coin. Many students, and even adults, think that if they get heads on one toss of a coin they will have a better chance of getting tails on the next toss. However, this is not true. The probability of getting heads or tails on each toss is exactly the same. In this lesson, for example, if a pair of dice were rolled 50 times and each time produced the "XX" (female) combination, the next roll still has an equal (50/50) chance of being male or female. This is a very difficult concept for some students to understand, and you may need to repeat it during the lesson.

When doing this activity, your students may notice that in a small population, such as 5 rolls of the dice, their results may not match the mathematically expected results of 50% males and 50% females. After the student groups have completed their 30 rolls of the dice, however, the combined results of the class should be closer to this 50/50 distribution of males and females. Extend this to the concept of an infinitely large population, in which the 50/50 results can be achieved.

Students usually learn the word "probability" in math class by the time they reach seventh grade. Before beginning the lesson, however, review the definition of probability with your students:

$$\text{probability} = \frac{\text{number of chances for an event}}{\text{number of possible events}}$$

Then relate probability to genetics as you discuss the handouts with your students. With most classes I have found that the only unfamiliar concept in this lesson is that of a gene pool.

Please note that this lesson may take more than one class period, depending on your class. Lesson 4 is a short lesson, however, so it can be combined with the spill-over time from Lesson 3.

Lesson Procedure

Step 1:

Divide the class into groups of two to four students. Distribute to each group one white die, one red die, one piece of felt (optional), one copy of Handout 1, "Population Worksheet," and one copy of Handout 2, "Gene Pool Summary Sheet," to each student.

Review the concepts learned in the first two lessons. Explain to students that they will be making a gene pool that includes dominant and recessive traits. Also explain how these genetic combinations are written (see the key for the chart).

Step 2:
Have students read the instructions on the "Population Worksheet." Check for understanding, then have students begin the activity. When each group has completed its population of 30 individuals, their results should be recorded on the "Gene Pool Summary Sheet." If necessary, allow students to use their handout from Lesson 1 as a guide in figuring percentages.

Step 3:
When all groups have completed their handouts, go over the genetic probability chart at the bottom of the "Gene Pool Summary Sheet." Make sure students understand the probabilities for male/female and dominant/recessive.

Step 4:
Compare each group's results by tallying them on the large chart you copied from the top of the "Gene Pool Summary Sheet." Discuss how these results compare with the mathematical probabilities listed on the bottom on the handout.

Step 5:
Discuss the writing assignment at the bottom of the "Gene Pool Summary Sheet," which can be done either in class or as homework. Make sure all students can define this lesson's key words, and that they understand the assignment. The class discussion and completion of the writing assignment will provide you with a way to check each student's mastery of the lesson.

Enrichment Activities

- ☐ Students can figure the probability of a certain number coming up on a die or the probability of a certain card appearing at the top of a card deck. This can be done at home or in class.
- ☐ Students can use the dice to indicate sex-linked genetic defects such as color blindness or hemophilia.
- ☐ Give students a hypothetical pedigree that they can use to tabulate the results for various generations of offspring.

Population Worksheet

Your group should have two dice—one red and one white. Each face of each die is marked D, d, X, or Y. Roll the dice several times until one X or Y shows face up on each die, and again until one D or d shows face up on each die. Record your results on the chart. The possible combinations are XX or XY; and DD, Dd, dD, or dd. Each trial is one individual in your population of 30.

Trial	D or d from red die (father)	D or d from white die (mother)	Genotype of offspring (DD, Dd, dD, or dd)	Sex of offspring (XY or XX)	Phenotype (use a descriptive word)
1					
2					
3					
4					
5					
6					
7					
8					
9					
10					
11					
12					
13					
14					
15					
16					
17					
18					
19					
20					
21					
22					
23					
24					
25					
26					
27					
28					
29					
30					

Gene Pool Summary Sheet

Population Summary

	Male	Female	Genotype	Phenotype
Total number				
Percentage				

Genetic Probabilities

Father Mother

Sex

XY — XX

XY or XX
Male or Female

50% each

Phenotype

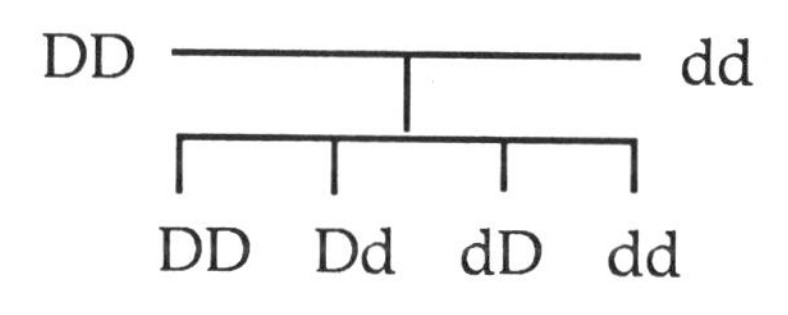

75% dominant phenotype

25% recessive phenotype

Write a paragraph comparing your totals with the genetic probabilities given above. Explain any difference between your results and the predicted genetic probabilities.

Lesson 4

T-Shirt Mania

Individual adaptation to an environment

Group Size
Individual students

Time Required
1 class period (or less)

Materials
- large sheets of colored butcher paper or construction paper in 5-6 colors
- copy for each student of both pages of Handout 1, "Hiding or Advertising?"
- extra T-shirts for those students who forget theirs (optional)

Key Terms
mutualism
predator/prey relationship
adaptation
environment

Instructional Goals
- To recognize that adaptation to the environment is critical for survival.
- To introduce the concepts of mutualism and predation as factors in adaptation.
- To recognize camouflage and advertising as strategies for adapting to the environment.

Student Objectives
Students will:
- Wear T-shirts to class that blend in or contrast with a known color that represents an environment.
- Observe which T-shirts are best adapted, through camouflage or advertising, to an environment.
- Tabulate their results by categorizing the T-shirts according to the adaptive relationship exhibited.

Prerequisite Knowledge
In addition to the material presented in the first three lessons of this unit, students should know that ecology is the study of the interaction between living organisms and their environment.

Advance Preparation Time
About 30 minutes
- Review the "Background Information" and "Lesson Procedure."
- Gather materials.
- Hang the sheets of colored paper around the room at various "stations," and number the sheets.
- Duplicate the student handout pages.
- Make a transparency of Handout 1, "Hiding or Advertising?" or copy the chart onto your chalkboard.

Teacher Tips
- Before the end of the previous class period, have students look at samples of the colored paper that will be used as the "environments" for this lesson. Students should decide whether they want to advertise (stand out) or be camouflaged (hidden) in one environment of their choosing.
- It is possible to do this lesson without giving students a choice of "hiding" or "advertising," but some students might feel uncomfortable if they stand out. Allowing students to make that decision the day before eases any qualms they might have.
- I suggest that you have several colored T-shirts on hand for those students who "forget" to wear them to class. You'll find, however, that most students like to have an excuse to wear a T-shirt to school.

Background Information

In this lesson students will manipulate three variables—coloration, environment, and interaction with another species. In addition, students will decide what kind of interaction occurs. Is the interaction an example of the *predator/ prey relationship* or *mutualism* via pollination?

Adaptation requires the use of different strategies under different conditions. It is an advantage to be well-camouflaged if an organism is trying to hide from a predator, as was the case with the Manchester peppered moth (see pages 79–81). However, if a flower needs a strategy to increase its chances for being pollinated, for example, the situation is quite different. Some plants depend on seed dispersal by the wind, others have seeds that cannot germinate without passing through a bird's intestines, and many plants rely on insects to carry pollen from one flower to another while they feed.

This lesson presents the concept of mutualism through the example of bee pollination. In order for the plant to benefit from the bee, it must be a good host. First, it must attract the bee's attention. The most obvious way is through its colorful flowers. Bees see ultraviolet light, which we do not see. To us ultraviolet light appears white. Bees also see the yellow/blue spectrum well. Thus a flower with a showy coloration may be attractive to a bee, but often the flower's ultraviolet center is what helps the bee find the pollen. Since a bee needs to be able to land on the flower to pollinate it, a plant needs a blossom with a landing pad—a strong stalk that holds the flower upright (such as clover), or petals that open up to support the weight of a bee. Tubular flowers, such as fuschia, are pollinated by hummingbirds; bees cannot land on these blossoms.

When discussing the predator/prey relationship with your students, be sure to point out that this relationship is also beneficial to the prey. Since many more offspring are born to the prey species than can survive, it is the predator that thins out the population of the weak and diseased. After all, they are the easiest to catch. This serves the prey population because defective individuals are removed from the gene pool, thus strengthening future generations.

Lesson Procedure

Step 1:
On the day before you teach this lesson, show students samples of the colored paper environments and ask them to wear a T-shirt to the next class that will either blend with or contrast with a particular environment. Number the environments (these numbers can then be referred to as station numbers).

Step 2:
At the beginning of the lesson, pass out your extra T-shirts to those students who forgot theirs. Distribute both pages of the student handout and go over the instructions with your students. Be sure to review the key terms.

Step 3:
Have students stand in front of their chosen environments, and ask them to fill out the chart on their handouts. Students will decide whether each person's T-shirt blends in with the environment (which is important in the predator/prey relationship), or contrasts with the environment (which is important in a mutualistic relationship).

Step 4:
After completing the handouts, have students return to their seats to compare results. With a transparency of the handout on the overhead projector, or a facsimile of the chart on your chalkboard, summarize the class responses.

Step 5:
After comparing results, have students list as many factors as they can that are involved in a predator/prey or mutualistic interaction. Tell them that color is just one factor. Give students a few moments to do this, and then share factors by calling on individual students. Write student responses on the chalkboard or on a blank transparency, and discuss other factors not mentioned by students.

Enrichment Activity

- Have students collect flowers (or pictures of flowers) and list each flower's attributes that might affect its chances for pollination—for example, coloration and petal formation. Students might research how flowers with different attributes are pollinated.

Hiding or Advertising?

Predator/Prey Relationship: Find a Hole and Hide!

In a snowy environment, which rabbit has the better chance for surviving and being able to reproduce?

If these moths landed on opposite trees, would they have a better chance of survival?

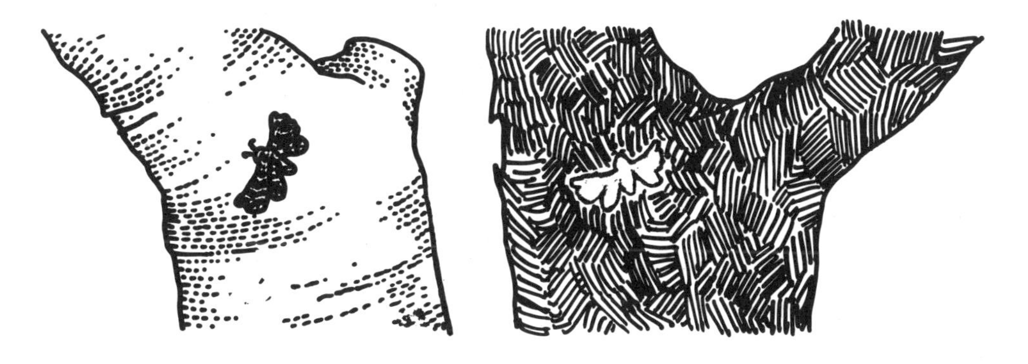

Mutualism: It Pays to Advertise!

If a blossom attracts the bee, it will be pollinated and produce seeds.

Both predator/prey relationships and mutualistic relationships contribute to natural selection of the gene pool.

Use this page to record your observations after each student stands in front of a colored paper "environment." Write the name and environment number of each student in the first two columns. In the third column write the color of each student's T-shirt. Write a check mark in the remaining columns to show whether each student is hidden or advertised, and which kind of adaptive relationship this represents.

Student	Environ-ment	T-shirt color	Hidden	Advertised	Predator/prey	Mutualism

The Birds and the Bees

Who will survive to produce offspring?

Group Size
2-4 students, using the same groups as in Lesson 3

Time Required
2-3 class periods

Materials
- 1 multicolored cloth remnant or rug sample for each group
- construction paper in colors to blend and contrast with the cloth or rug samples—3 sheets of each color
- paper hole puncher
- small containers (such as plastic bowls or cups) to hold dots punched out of the construction paper;
  1 container for each color
- 2 muffin tins for sorting
- plastic sandwich bags for storage of each color of dots
- 1 timer or stopwatch (a watch with a second hand works fine). If you have each group time the activity themselves, then each group will need a watch or timer.
- copy for each student of Handout 1, "Camouflage: Who Survives?" and Handout 2, "Advertising for Pollinators"
- copy for each group of Handouts 3 and 4, the "Tally Sheets"

For optional activity:
- 1 pair of lab goggles for each group
- several sheets of transparent plastic in various colors
- scissors, tape, marking pens

Key Terms
natural selection
survival of the fittest
dominant/recessive
predator/prey relationship
mutualism
adaptation

Instructional Goal
- To synthesize the previous four lessons by having students use the population they formed in Lesson 3, give it a dominant/recessive color trait, and place the "individuals" in an environment. Students will discover which individuals are best suited to the environment in a way that predators will have difficulty finding them and pollinators will be able to sight them.

Student Objectives
Students will:
- Place a population in an environment to determine the individuals that are best adapted for camouflage and best adapted to advertise.
- Role play as predator and/or pollinator, selecting those individuals that are not adapted to survive.
- Form a new simulated population by mating the survivors.
- Predict which traits are best adapted and check that prediction.
- Formulate variables that might affect natural selection.
- Write an analysis of the Manchester peppered moth story, which was presented on pages 79–81.

Prerequisite Knowledge
Students should be familiar with the material in Lessons 1-4.

Advance Preparation Time
1-3 hours
- Review the "Background Information" and "Lesson Procedure."
- Gather materials.

- Have students help punch dots out of the colored construction paper. Make sure the collection tray on the paper hole puncher is cleaned out before starting, and after each color change, so the colored dots aren't mixed. As the dots are punched, collect them in the muffin tins to keep each color separated. When all the dots have been sorted, put each color into its own container, such as a small plastic bowl or cup. After the activity is over, store the colored dots in plastic sandwich bags for use next year.
- Number each cloth or rug environment.
- Duplicate the student handouts.
- Make a transparency of the "Class Summary Chart" on page 109, or write this chart on the chalkboard.

Teacher Tips

- Be sure to go over the instructions on Handouts 1 and 2 with your students. Each student should understand the purpose of the experiment and should remember the key terms from Lesson 3—especially genotype and phenotype—as well as how dominant and recessive traits are paired and notated. When teaching this lesson, I review these terms and concepts while going over the instructions.
- To insure consistency I have found it works best to time the 30-second "predation period" yourself, while the whole class does the activity at the same time.
- If none or only a few dots remain on the cloth or rug environment after the first trial, repeat the experiment for a shorter time period.
- As an alternative to colored paper dots, you might be able to find candy confetti dots in colors appropriate for your cloth or rug environments. This way the student "predator" can actually eat the prey. This requires a nominal added expense, and it can be somewhat tedious to separate the confetti colors.

Background Information

In the previous four lessons, students learned about variation, dominant and recessive traits, and phenotype and genotype. They created a hypothetical population and discovered how individuals avoid predators or attract pollinators. In this lesson, students synthesize this knowledge in an activity that includes all these concepts and shows exactly what happens to two generations of the gene pool. Only those individuals that are left after the predator has fed, or those that were selected by the pollinator, will be able to breed and produce a new generation. This activity allows students to observe, in a concrete way, how the selection process works.

The only new concept emphasized in this lesson is that of *survival of the fittest*. You'll find that many students think survival of the fittest means that individuals with "good" attributes, like strength or large size, are the only ones who survive. Be sure to point out that, in evolution, survival of the fittest means that those individuals best adapted to their environment not only survive, but are also able to reproduce. If an individual animal survives birth, but dies before it is able to breed, its genetic make-up is lost to the gene pool.

Since it takes two individuals of the opposite sex to produce an offspring (in sexual reproduction), the number of males and females in the gene pool is just as important as what traits exist in the pool. After all, if only one male is left after a predator has fed, only that male will be able to mate. His genetic make-up will become the most prevalent in the next generation, no matter what characteristics are present in the females.

Most students enjoy this lesson and like to ask "What would happen if ..." questions. It's easy to extend this lesson to a third period and let your students test their hypotheses. See the "Optional" steps under "Lesson Procedure For Day 2" for more information.

Lesson Procedure For Day 1

Step 1:
Divide your class into the same student groups as in Lesson 3, and have students use their "Population Worksheets" from that lesson. Distribute Handout 1 to each student and Handout 3 to each group. Give each group a numbered cloth or rug environment.

Step 2:
Go over the instructions on Handout 1 with your students and monitor their work for the first three steps. First, students need to decide which trait (color) will be dominant and which trait will be recessive. Students should then choose who will be the predator, recorder, and timer (optional). Point out they will have a chance to exchange roles. Be sure the "predators" turn around while other group members place their dots on the cloth or rug environments.

Step 3:
When all the groups are ready, start timing the 30-second predation period, or have one student in each group start timing. Student predators then remove dots, one at a time, from the environments. When role playing as predators, students should try to remove the dots in the order in which they see them.

After the time limit is up, students should count the number of dots of each color that are left in the environment. The recorder marks off the "deaths" on the original "Population Worksheet" by placing an "X" next to the appropriate phenotype column. Each group's recorder then writes the color (phenotype), genotype, and sex of the survivors on Handout 3, "Tally Sheet 1," under Trial 1.

Step 4:
Student groups then repeat the experiment with the surviving members of the population. Results for this second generation are recorded on "Tally Sheet 1" under Trial 2.

Have each group give you their totals for both trials and enter the results on the "Class Summary Chart" (see Figure 1 on page 109) on the chalkboard or overhead projector. Your students should be able to see some definite patterns emerge. They will know which color is best adapted to its environment and should be able to predict what might happen if they repeated this experiment for another generation. Discuss these predictions and have students suggest other factors besides the choice of color and placement on the environment that might affect the outcome of the experiment. This will prepare them for their homework assignment, which is presented at the bottom of Handout 1. Students shouldn't find it difficult to come up with suggestions, such as the eyesight of the predator, the taste of the prey, etc.

Then have each group return the containers of dots and their cloth or rug environment.

For Day 2

Step 1:
Distribute copies of Handout 2, "Advertising for Pollinators," and Handout 4, "Tally Sheet 2." Each group should use the same cloth or rug environment as on Day 1, but this time the groups should choose two colors that *contrast* with the environments. Then each group should decide which color will be dominant, which color will be recessive, and who will be the "pollinator."

Step 2:
The experiment is repeated (2 trials), and each group should record their results on the population and tally sheets as before. This time the dots removed by the pollinator will be the survivors, since they have been pollinated. Then record all the group results on the Class Summary Chart.

Discuss the results. Have students share their suggestions from the homework assignment and think of more new variables. After this experiment, they may suggest variables such as flower shape, sugar content of the pollen, etc. This discussion can lead to an optional experiment (see below) if you have an enthusiastic class and extra time.

Step 3:
To introduce the homework assignment at the bottom of Handout 2, review and discuss the Manchester peppered moth story (pages 79–81) with your class. Have your students write an analysis of the story, explaining how it demonstrates the concepts they have learned in this unit—variation, dominant/recessive traits, predator/prey relationships, adaptation, gene pool, natural selection, and survival of the fittest.

I use this assignment as a major portion of each student's grade for this unit. Since I also teach English, I grade the assignment on writing mechanics as well as on science content. I usually give students several days to complete this analysis, with a first draft or outline due after one day. If students are having problems, you will find out before they have spent too much time on the wrong tack.

Optional:
The same experiments can be performed using variables your students have suggested. One activity I've used involves the variable of eyesight. If your students determine that the eyesight of the predator and/or the pollinator may affect the results of the experiment, they can tape different colored pieces of transparent plastic to a pair of lab goggles, simulating this variable (see page 105 for a list of suggested materials). Likewise, if students decide taste might be a factor, certain dots could be marked to indicate bad taste. The variables that can be experimented with are limited only by your students' imaginations.

Enrichment Activities

- ☐ Add a lesson on mutations as a variable in the gene pool. Sometimes students will think of this suggestion on their own.
- ☐ Have your students observe and describe an environment, such as their backyards or a local park, and an organism that is well-adapted to this environment.
- ☐ Students can find an example of how pollution or human interference disturbs the natural selection process of a population. Research can be done using newspapers, magazines, or TV programs, in addition to library resources.

Figure 1. Class Summary Chart

Camouflage: Who Survives?

Trial 1—1st Generation

Environment number	Number with dominant phenotype	Number with recessive phenotype

Trial 2—2nd Generation

Environment number	Number with dominant phenotype	Number with recessive phenotype

Advertising for Pollinators

Trial 1—1st Generation

Environment number	Number with dominant phenotype	Number with recessive phenotype

Trial 2—2nd Generation

Environment number	Number with dominant phenotype	Number with recessive phenotype

Camouflage: Who Survives?

Procedure:

1. Have one member from your group get 1 multicolored cloth or rug "environment" and 2 containers of dots with colors that blend in with the environment. These colors represent dominant and recessive traits in a gene pool.

2. Using your "Population Worksheet" from Lesson 3, you will form a gene pool containing a dominant and a recessive color. One of the dot colors will represent a dominant trait and the other will represent a recessive trait. The choice is yours.

   ___________ colored dots are dominant

   ___________ colored dots are recessive

   The dominant color you chose will be the same as the *phenotype* on your "Population Worksheet." Remember, all genotypes of DD, Dd, or dD will be the dominant phenotype. Only the genotype dd can be the recessive phenotype.

   Which color to you think will be best camouflaged in your environment? ___________

3. Take out from each container the number of colored dots that corresponds to each phenotype on your "Population Worksheet."

   For example, if you chose tan as dominant and white as recessive, you would use the same number of tan dots as you have dominant phenotypes, and the same number of white dots as recessive phenotypes. This is your gene pool.

4. Decide which group member will be the "predator." (You will be able to take turns being the predator.) Have the predator turn around while the rest of the group members place the dots on the cloth or rug environment. Since this is an experiment about the predator/prey relationship, try to place the dots so they blend in well with the environment, as if they were hiding from a predator.

5. The "predator" will have 30 seconds to "feed" on the dots by removing them one at a time. The removed dots represent the number of prey eaten by the predator. They did not survive to reproduce, so their genetic make-up has been removed from the gene pool.

6. Record the number of "dead" dots by going to the top of the "phenotype" column on your "Population Worksheet" and writing an "X" to the right of the phenotype that corresponds to each removed dot.

Those phenotypes without an "X" are the survivors. Transfer each survivor's sex, genotype, and phenotype to your "Tally Sheet" under Trial 1. Write in the colors you chose under phenotype. This is your new gene pool for the second generation.

7. Using the results of your first trial, answer the following questions:

   Which color was best adapted to your environment?

   ____

   Was the best-adapted color dominant or recessive?

   ____

   Was your original hypothesis (in question 2) correct?

   ____

8. Using the survivors of your first trial, place the dots back on the cloth or rug environment. Then repeat the experiment for the second trial (second generation). This time have another student play the role of predator.

9. After 30 seconds count the survivors and record the results under Trial 2 on your "Tally Sheet." Give your results for Trials 1 and 2 to your teacher so they can be written on the Class Summary Chart. Then return your environment and containers of dots.

Homework Assignment
Design an experiment similar to the one you did in class with a new variable that might affect the results. Be sure to list the experiment's procedures, as well as any materials needed. Use this handout as a model if you like.

Advertising for Pollinators

Procedure:

1. Use the same cloth or rug environment you used before. Then get 2 containers of dots with colors that *contrast* with the environment. This time instead of having dots that will be camouflaged, you want them to stand out against the environment so a pollinator will be attracted to them.

2. Decide which color will be dominant:

   ____

   Which color will be recessive?

   ____

   Which color do you think will advertise best?

   ____

3. A group member who did not get to act as a predator on the first day should be the pollinator for this activity. Have the pollinator turn around while the rest of the group places the dots so they will be good advertisers.

4. As before, the pollinator has a 30-second time limit to remove dots one at a time. After the time limit is up, count the number and color of dots removed. These are the survivors—they have been pollinated.

5. Using your "Population Worksheet" from Lesson 3, start at the top and put an "X" next to the phenotypes of the dots that were removed. Write the genotype and phenotype of the individuals with an "X" on your Tally Sheet under Trial 1. This time we are not concerned with the sex of the individuals because the bee is responsible for the pollinating.

6. Using only the removed dots (pollinated individuals), repeat the activity and record your results under Trial 2 on your Tally Sheet, then answer the following questions:

   Which color was best adapted to its environment?

   ____

   Was your original hypothesis (in question 2) correct?

   ____

   You took turns in playing the role of pollinator and predator in order to be fair. However, this procedure might not be considered a good control. Why not?

   ____

   ____

   ____

If you had known beforehand that bees see ultraviolet light, which appears white to us, but *do not* see the red spectrum, how would this have affected your results?

____

____

____

____

Homework Assignment
Write an analysis of the Manchester peppered moth story. Explain how this true story demonstrates the concepts you learned in this unit. Be sure to use these key terms:

- variation
- natural selection
- survival of the fittest
- predator/prey relationship
- dominant/recessive traits
- gene pool
- adaptation

Tally Sheet 1

Camouflage: Predator/prey relationship

Dominant Color: ______________ Recessive Color: ______________

Trial 1	Sex	Genotype	Phenotype
1			
2			
3			
4			
5			
6			
7			
8			
9			
10			
11			
12			
13			
14			
15			

Total Dominant Phenotype: ______

Total Recessive Phenotype: ______

Trial 2	Sex	Genotype	Phenotype
1			
2			
3			
4			
5			
6			
7			
8			
9			
10			
11			
12			
13			
14			
15			

Total Dominant Phenotype: ______

Total Recessive Phenotype: ______

Tally Sheet 2

Advertising: Mutualism

Dominant Color: ____________ Recessive Color: ____________

Trial 1	Genotype	Phenotype
1		
2		
3		
4		
5		
6		
7		
8		
9		
10		
11		
12		
13		
14		
15		

Total Dominant Phenotype: ______

Total Recessive Phenotype: ______

Trial 2	Genotype	Phenotype
1		
2		
3		
4		
5		
6		
7		
8		
9		
10		
11		
12		
13		
14		
15		

Total Dominant Phenotype: ______

Total Recessive Phenotype: ______

Ecosystems
Nature's Balancing Act

Tom Wight
Rolling Hills Middle School
Watsonville, California

I entered the teaching profession in 1959 as an agricultural science teacher in the small farming community of LeGrand, California. At first I was a model by-the-book teacher. I conducted my classroom according to the rather rigid concepts I had been taught during my formal training period, resulting in a friendly but distant relationship with my students and their families. After two years, I left the teaching profession for a job in sales that offered a higher salary than teaching. Three years later, after developing a bad case of ulcers from the constant pressure of meeting ever-expanding sales quotas, I went back to the classroom as a seventh grade teacher in Monterey County. An eighth grade teacher and I agreed that he would teach reading and English to my students and I would teach science and math to his students. This arrangement allowed me to adopt a teaching style that was more suited to my personality, and I began to enjoy what I was doing.

After four years I accepted a position as a self-contained fifth and sixth grade teacher at Amesti Elementary School, where I remained until last year. Concurrently I earned a master's degree in education from Notre Dame College in Belmont, California, with an emphasis in outdoor and environmental education. As a result of this training, I expanded my curriculum to include week-long field trips to Yosemite and participation in the Santa Cruz Outdoor School Program. I now teach at Rolling Hills Middle School, and I truly look forward to entering my classroom each day—the antithesis of my first two years of teaching.

Lessons:

1. "The Freshwater Aquarium"
2. "Adaptation in Aquatic Environments"
3. "Effects of Pollution on an Ecosystem"
4. "Interdependence"

Overview

EARTH HAS OFTEN BEEN REFERRED to as a planetary spaceship, since it has only a finite amount of land, water, and air to support all the life forms found here. Approximately three-fourths of Earth's surface is covered with water. No living organism can survive for long without this wondrous substance that contains only two atoms of hydrogen and one atom of oxygen. We even carry a "miniature sea" inside our bodies that is remarkably similar in content to the saltwater found in the oceans.

The quality of the environment and the continued existence of healthy life forms are interdependent. Through technological expertise, greed, and a seemingly insatiable desire to subjugate all of nature, humans are the only animals that have the capacity to seriously alter, and even obliterate, the global environment.

Because water is so vital to the health and welfare of every living thing on this planet, we need to carefully examine what we are doing to this precious substance in terms of the extensive pollution we are producing in ever-increasing amounts. I am told by people who spend their lives on the sea that evidence of this pollution goes floating by their vessels every day in the form of garbage or globs of oil. Lakes and forests all over the world are dying from the effects of acid rain. Man-made monuments are not spared either. Buildings and statues are gradually being eaten away by the corrosive action of acid rain. In many places domestic and community wells have had to be sealed because dangerous industrial chemicals have infiltrated the aquifers beneath them. Birth defects and certain types of cancer have been traced directly to polluted water sources.

We need to teach our children early in their educational experience to develop an ethical attitude toward our environment and concern for its limited planetary resources. As science educators, we have the great responsibility of making children aware of the fact that water, especially freshwater, is a finite resource and that each of us has the duty to protect its quality. It is my hope that by teaching the lessons in this unit, which use simple aquatic environments, you will begin to achieve this goal with your students.

Key Concepts

- ☐ Plants and animals are interdependent in a balanced ecosystem.
- ☐ Populations may stabilize over a period of time in a balanced ecosystem.
- ☐ Producers, consumers, and decomposers each play an important role in an ecosystem.
- ☐ Food chains and food webs indicate the flow of energy that maintains habitats within an ecosystem.
- ☐ Adaptation enables organisms to live in their particular environment.
- ☐ Various kinds of pollution have an effect on ecosystems.

Skills Used in the Lessons

- observing
- predicting
- communicating
- comparing
- organizing
- relating
- applying
- inferring
- concluding

Extensions and Sources

Books I used as references for setting up an aquarium ecosystem are listed below.

Cohen, Sylvan. *Enjoy Your Aquarium.* New York: The Pet Library Ltd. (Usually available in pet stores and supermarkets.)

Elementary Science Study. *Teacher's Guide to Brine Shrimp: Observing the Life Cycle of a Small Crustacean.* New York: McGraw-Hill, 1975.

Minnesota Environmental Sciences Foundation, Inc. *Brine Shrimp and Their Habitat.* Washington, D.C.: National Wildlife Federation, 1972.

Outdoor Biology Instructional Strategies: Water Breathers. Berkeley, CA: Lawrence Hall of Science, University of California.

Strongin, Herb. *Science on a Shoestring.* Menlo Park, CA: Addison-Wesley, 1985.

I have found that the following resources are practically indispensable for teaching environmental and ecological concepts.

Cornell, Joseph Bharat. *Sharing Nature with Children.* Nevada City, CA: Ananda Publications, 1983.

Humboldt Environmental Education Project. *Green Box.* Eureka, CA: Humboldt County Schools, 1975. (Write to Humboldt Environmental Education Project, Humboldt County Schools, Office of Environmental Education, Sixth and H Streets, Eureka, CA 95501.)

Jorgensen, Eric, et al. *Manure, Meadows and Milkshakes.* Los Altos, CA: Hidden Villa Trust, 1986. (Write to Hidden Villa, Inc., 26870 Moody Road, Los Altos Hills, CA 94022.)

Kolb, James. *Marine Science Project: For Sea.* (Write to Marine Science Center/ESD 114, 17771 Fjord Dr., NE, Poulsbo, WA 98370.)

Project WILD. (Manual and workshops available through the National Wildlife Federation, 1412 Sixteenth St., NW, Washington, D.C. 20036.)

Van Matre, Steve. *Acclimatization.* American Camping Association, 1972. (Especially helpful if you are interested in a sensory and conceptual approach to ecological involvement.)

Glossary

Biome: A large geographical region made up of a group of ecosystems in which climatic factors usually play a role. The two major types of biomes are aquatic and terrestrial.

Community: Interacting populations of different plant or animal species that live in the same area. A redwood grove may have a preponderance of redwoods living there, but squirrels, insects, birds, and other life forms will also occupy the grove.

Consumer: An organism that uses other organisms—plant or animal—for food.

Decomposer: An organism that obtains its food from dead organic matter.

Ecology: The study of organisms and their relationship to and interaction with the environment.

Ecosystem: A community of living things interacting with each other and with the physical environment. The populations living in an ecosystem affect each other and are affected by the physical aspects of the environment. The physical environment will eventually be affected by all the various species living there. For example, the soil in the redwood grove will be enriched by decaying plant and animal material. As long as a balance exists between the living and non-living factors, and barring some catastrophic event such as a fire or flood, an ecosystem will generally do a good job of taking care of itself over a fairly long period of time.

Environment: In a biological sense, this word means the aggregate of surrounding things, conditions, or influences—biotic and abiotic factors—that are required by each species to maintain life. The major abiotic factors in an environment include such things as light, water, temperature, soil, oxygen, and minerals. The main biotic factors in an environment generally include such things as predators, prey, plants, and animals. Several types of relationships between organisms exist in an environment: *symbiosis,* in which organisms live together for mutual benefit; *commensalism,* in which organisms live together in a relationship where only one benefits but the other is not harmed; and *parasitism,* in which organisms live together in a relationship where one organism is harmed and the other one benefits.

Food chain: The transfer of food energy from the source in microscopic plants through a series of animals.

Food web: An interlocking pattern of food chains.

Habitat: The specific area in the environment where a particular species makes its home and has the most satisfactory interrelationship with its environment.

Interdependence: The mutual dependency between organisms for their continued survival.

Population: A group of plants or animals of the same species inhabiting a particular region. For example, all the redwood trees living in a particular grove make up a population.

Producer: A plant in a community that manufactures food; it is generally referred to as a primary food producer.

Lesson 1

The Freshwater Aquarium

A mini-ecosystem for classroom study

Group Size
Individual students, or groups of 4

Time Required
1 class period for aquarium set-up;
then daily observation for 4-12 weeks

Materials
- 1 five- or ten-gallon plastic aquarium
- clean coarse sand or fine gravel
- under-gravel aquarium filter
- air pump and tubing
- aquarium heater with thermostat
- pH kit, sodium biphosphate, sodium bicarbonate
- sponge aquarium scraper
- glass cover
- aquarium light
- aquarium thermometer
- aquarium net (small to medium size)
- several sprigs of aquatic plants, either *Elodea* or *Anacharis*
- fish (guppies or goldfish—see "Lesson Procedure," pages 123–125)
- 1 or 2 aquatic snails (*Ampullaria cuprina*)
- aged (dechlorinated) water
- copy of Handout 1, "Making a Fresh-water Aquarium," for each student
- each student should have a journal for recording observations

Note: Aquariums, plants, and fish can be purchased at a pet shop or through science equipment suppliers.

Key Terms

population	producer
community	consumer
ecosystem	decomposer
environment	food chain
habitat	food web

Instructional Goals
- To demonstrate that producers, consumers, and decomposers each play an important role in an ecosystem.
- To demonstrate that adaptation enables organisms to live in their particular environment.
- To demonstrate that food chains and food webs indicate the flow of energy within an ecosystem.
- To demonstrate that populations may stabilize over a period of time in a balanced ecosystem.
- To demonstrate that plants and animals are interdependent in a balanced ecosystem.

Student Objectives
Students will:
- Make a prediction (hypothesis) regarding the characteristics and requirements of a balanced ecosystem.
- Help construct and maintain an aquarium.
- Observe that aquatic organisms interact with and adapt to their environment.
- Observe the territorial behavior of aqautic animals.
- Identify the role of producers (plants), consumers (fish and snails), and decomposers (bacteria and snails) in an aquatic environment.
- Observe food chains and webs in an aquatic environment and relate them to their energy source—the sun.

Prerequisite Knowledge
Students should understand the key terms, as well as the basic concept of an ecosystem and the processes that contribute to its functioning.

Advance Preparation Time
About 3 hours
- Review the "Background Information" and "Lesson Procedure."
- Gather materials.
- Duplicate the student handouts.
- Decide on a location for your aquarium. It should be placed away from direct sunlight to prevent the growth of excess algae.
- The day before you assemble the aquarium in class, fill a clean container with the amount of water you will put in the aquarium. The water must stand for at least 24 hours before you use it in the aquarium so that the chlorine will escape into the atmosphere.

Teacher Tips
- As you read the list of materials needed for this activity you may be saying to yourself, "I don't have time to gather all these pieces of equipment," or, "I don't have enough money in my budget to make the necessary purchases." I had exactly the same concerns when I contemplated adding an aquarium to my science classroom a few years ago. I decided to write a letter to the parents of my students to see if they could help me out. The response was overwhelming. I didn't get just one aquarium, but five aquariums and all the related equipment—air pumps, filters, tubing, the works. In addition to these things I was also given a microscope, an iguana and its cage, a boa constrictor and its cage, two guinea pigs and their cages, and a set of metric scales. I think we sometimes forget that the parents of our students are a resource that can be tapped. It is amazing how much stuff families have tucked away on a closet shelf or in the garage that they or their children no longer want. I give a receipt for all donated items so that those parents who itemize have a tax deduction.
- Aquarium tanks come in an almost endless variety of sizes and shapes. I recommend that you purchase a rectangular plastic five- or ten-gallon tank for two reasons: they are virtually indestructible, and they have no metal parts to corrode.
- I prefer that each group of students has its own small aquarium. I am a strong believer in student ownership of science projects. They are far more likely to take good care of the projects they have constructed than they are to take care of the one that I have put together. Also, it is far easier for four students to observe an aquarium than it is for a whole class to observe one. I recommend one aquarium for four students, since space is limited in most classrooms.

  For these student aquariums, I use one-gallon glass jars with wide mouths, such as mayonnaise jars used in school cafeterias or restaurants. You can find round undergravel filters, available in many pet stores, that will just fit inside these jars. The procedure for setting up these aquariums is the same as for the larger classroom aquarium, which is described on pages 123–125.

  If budgetary restrictions are of concern to you, as they are to me, you can purchase a single heavy-duty air pump that will drive the filters in several of these small tanks. I have one large air pump that drives the filter systems in 12 one-gallon aquariums. A single air line is attached to the pump, and brass fittings are evenly spaced along this line. The brass fittings permit me to attach each of the small aquarium filter systems to the air supply. Each fitting has an adjustable valve that controls how much air is delivered to each filter.
- I rarely use goldfish in my classroom aquarium. I prefer guppies for a variety of reasons—they are inexpensive, they reproduce in a short period of time, they give birth

to live young, they don't dredge up the gravel at the bottom of the tank, the males are easy to distinguish from the females, and they are easy to feed and maintain.

- In my experience, exotic fish should not be used in the average classroom. These fish tend to be more expensive and are not easy to maintain. However, if you would like to tackle this challenge as an extension of this lesson, get a parent who keeps tropical fish as a hobby to come in, explain, and set up this type of aquarium with student aides.
- In my classroom I maintain a second ten-gallon aquarium as a "mother" tank from which I can extract plants, fish, or water for various science activities.
- When you purchase the air pump, don't settle for the cheapest one. In my experience, it's better to buy a good quality pump that will last for several years.

Background Information

What is an aquarium? The dictionary defines an aquarium as a pond, tank, bowl, or the like, in which living plants and/or animals are kept for exhibit or study. That is a good definition as far as it goes. I would like to add that an aquarium is one of the most interesting, motivating, and useful tools that you can create for teaching science in your classroom.

In my opinion, a science classroom without an aquarium is like an art museum without any art. Children, and many adults I know, love to watch fish and other aquatic creatures in an aquarium. There is something very calming about this activity. You may have noticed that many doctors' and dentists' offices have aquariums in their waiting rooms, and for good reason. Aquarium-watching has a measurable effect on reducing stress. In addition to being interesting and reducing stress, an aquarium is a perfect tool for you to use in teaching many different biology and physical science lessons to your students.

The actual construction of a classroom aquarium takes only one class period. I build the aquarium *with* my students, explaining each step as I go along and allowing them to do as much as possible. To accomplish the other student objectives will take a longer period of time. I generally allow four to twelve weeks for students to observe the changes in the aquarium ecosystem.

You should set aside some time each day for your students to observe the aquarium and record their findings in their journals. Nothing startling generally happens from day to day; most of the phenomena taking place in an aquarium are subtle and will require careful observation on the part of your students. Things students should observe include changes in the population of plants, fish, or snails; water temperature; algae and plant growth; water pH; and clarity and odor of the water. Some of this data are more easily kept in graph form, such as water temperature and population figures. It isn't necessary to check the pH of the water every day, but once a week your students should record the pH in their journals. They usually enjoy watching the color develop on the litmus paper that indicates the pH of the water.

During the weeks of aquarium observation, introduce your students to other types of ecosystems and the principle that a balance of biotic and abiotic factors is required for an ecosystem to function successfully. If you can, include a field trip to a public aquarium or nature preserve to observe other ecosystems.

Lesson Procedure

Step 1:
When you first bring the aquarium tank into your classroom you will probably want to rinse it out. However, *do not* wash the aquarium or any of its associated equipment with soap or detergents. No matter how hard you try you will be unable to remove all the soap residue. Aquatic life forms are very sensitive to the harmful chemicals presents in these products. If you cannot adequately clean the tank with plain hot water, I recommend scrubbing it with baking soda.

Ask your students to hypothesize about what equipment might be necessary to put in an aquarium so it can maintain a balanced ecosystem. Then distribute the student handouts.

Step 2:
After the aquarium has been cleaned, place an under-gravel filter at the bottom of the tank and attach the air line tubing to the appropriate part of the filter. Some of these filters come with an activated charcoal component, but in a balanced aquarium this is not really necessary and may even interfere with some of the experiments your students may want to perform. (*Note:* An aquarium filter is not absolutely necessary. Without a filter, however, the water may become cloudy over a period of time because of the growth of bacteria and fungi. Without a filter you will have to change the water more frequently to keep it from becoming polluted.)

Step 3:
Place the coarse sand or fine gravel in a large kitchen strainer and run tap water through it to remove dust and dirt particles. If you eliminate this step, all the dirt and dust in the sand will muddy the water and/or rise to the surface of the tank, resulting in a mess. Sand is only necessary if you are keeping guppies or if you are using an under-gravel filter. However, if you do not use sand you will need to change the water more frequently. Sand provides a place for bacteria to grow and feed on dead plant and fecal matter.

Set the aquarium in the location you have designated for it (away from direct sunlight), and cover the under-gravel filter with 1-1/2 to 2 inches of the clean sand or gravel.

Step 4:
Fill the tank with aged water. This is water that has been allowed to stand for at least 24 hours so that any chlorine or other purifying chemicals in the water have time to escape into the atmosphere. Most aquatic organisms are very sensitive to chlorine and may even die when exposed to it.

Mark the exact level of the water on the outside of the tank with a piece of masking tape. As the water in the tank evaporates, the minerals in the water are left behind and become more concentrated. To maintain a balanced environment it is necessary to keep a constant level of water in the tank. Remember to add only aged water.

Step 5:
Attach the other end of the air line tubing to the air pump and turn on the pump. Bubbles of air should begin to rise from the filter's tube to the surface of the water. As this action continues, even though you cannot observe it, water will begin to move through the gravel filter and out the filter's tube.

Step 6:
Now you can place aquatic plants in the aquarium. I generally add the plants after the aquarium has been filled because it eliminates the problem of how to put water in the tank without uprooting the plants. *Elodea* and *Anacharis* are both good plants for an aquarium. Both plants are rootless but need to be anchored in the gravel.

A small rock can be used for this purpose. The plants will grow toward the water's surface, eventually floating there. *Anacharis* can grow very quickly, up to an inch a day when established, and may occasionally need to be trimmed. Plants are necessary in an aquarium to help provide the oxygen needed by the fish and snails, to use the carbon dioxide that the fish and snails put in the water through respiration, and to provide some of the food for the fish, depending on the type of fish you place in the aquarium. I do not recommend you use plants as the sole source of food for the fish. (See Figure 1, which shows a sample aquarium set-up.)

Step 7:
I generally wait a day or two after setting up the aquarium before I put the fish into the tank. This allows time for any cloudiness in the water to settle. *Caution:* Do not put the fish directly into the aquarium. Most fish are very sensitive to sudden changes in water temperature, and I have seen fish die within moments after being suddenly placed in water of a different temperature. To avoid this problem, put the fish into a plastic bag filled with water from the tank in which they were taken. Then float the bag on the surface of the new tank for approximately one-half hour. The water temperature in the bag will slowly adjust to the water temperature in the tank, giving the fish a better chance of adjusting to their new environment. Then add a little water from the tank into the bag. After completing this process, the fish can be safely released in their new home. The snails can be placed in the aquarium at the same time you add the fish. One or two snails for a five-gallon aquarium is plenty. Like guppies, snails reproduce at a phenomenal rate. Soon you will notice gelatinous masses of eggs on the aquarium walls at or above the surface of the water. Within a couple of weeks you and your students can watch the eggs hatch. Snails eat algae primarily—the green film that grows on the aquarium walls. Remember that the amount of light your aquarium receives has a direct bearing on the growth of algae. A balanced aquarium ecosystem will not produce more algae than the snails can consume. *Ampullaria cuprina,* the species of snail I recommend for your aquarium, also eats such things as dead plant matter and leftover fish food, but generally will not eat the living plants.

The number of fish and snails you place in the tank will have a direct bearing on the quality of the aquarium's environment. If you put too many fish in a tank, especially goldfish, you will have a real ecological disaster on your

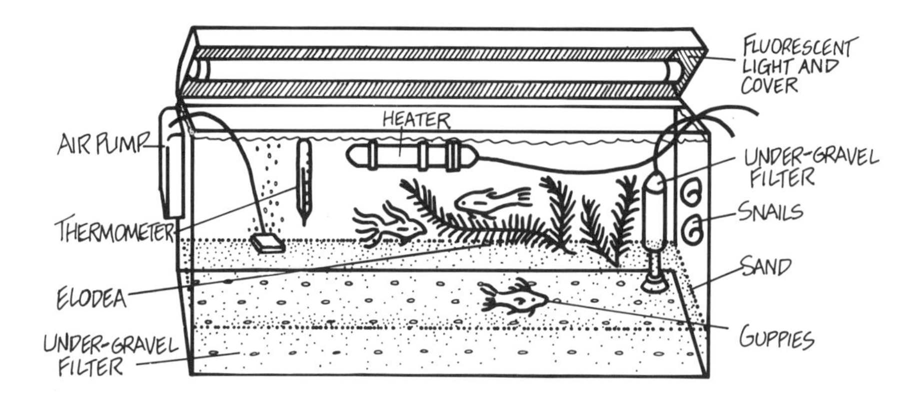

Figure 1. A sample classroom aquarium. Your equipment may differ in appearance or placement in the aquarium, but you should obtain all these components.

hands. (Of course, this may be the result you are looking for if your students are conducting an experiment on the effects of overpopulation.) The number of fish you should place in an aquarium is directly related to the volume of water in the tank and the amount of oxygen available. If you are using guppies, I have found that one or two adult males and four to six adult females per gallon of water is plenty to start with because of their reproductive capacity. Two to three adult goldfish per gallon seems to work well. If I use goldfish, I buy the "feeder" variety because they are not so expensive and they are smaller than the other types. A rule of thumb is an inch of fish for every four to five square inches of water surface. A good indication that there are too many fish in the tank is when you notice fish at the surface seemingly "gasping" for air. Fish can drown when there is not enough dissolved oxygen in the water.

Step 8:
Left alone, an aquarium will slowly take on the temperature of the surrounding environment. In my experience goldfish are not too fussy about water temperature. Guppies, on the other hand, only do well with the water temperature between 60 and 72 degrees Fahrenheit. During the day, most classroom ambient temperatures are about 68 to 72 degrees, which is fine for guppies. However, if your school is like mine, your space-heating system is controlled by the clock and the temperature drops overnight and on weekends. Water does not readily give up its heat, but eventually an aquarium will cool down to the lower temperature of your classroom. Over a weekend it can suffer a substantial loss of heat. Therefore, if you are raising guppies, I recommend you use a thermostatically-controlled aquarium heater. This will extend the lifespans of your guppies and contribute to the success of your aquarium ecosystem.

Step 9:
Once a week or so, have students discuss the observations and findings they have written in their journals. They can hypothesize about the reasons for changes, as well as predict changes that may occur in the aquarium ecosystem. I use the following questions in class discussion to check for understanding.

- What changes have occurred in the aquarium?
- Why do you think these changes have taken place?
- What was the average temperature of the aquarium?
- What is the pH of the aquarium water?
- What is pH and why is it an important abiotic factor in the aquarium?
- What happens to the waste material produced by the plants and animals?
- Why are animals like snails valuable in an aquatic environment?
- How is an aquarium similar to a lake or pond?
- If we were to construct an outdoor pond, what factors would we need to take into consideration?
- Describe one food chain you have observed in the aquarium and compare it to another food chain outside the aquarium. How are these two food chains similar and how are they different?
- What is a producer? A consumer? A decomposer?
- What is a balanced ecosystem? Is our aquarium balanced?

Step 10:
Students can design and implement their own experiments to test variables within the aquarium ecosystem. Be sure they follow the steps of the scientific method and obtain your approval before they start. Students should continue to write their observations in their journals. Depending on the experiment, results may be visible immediately or may take several weeks.

Management Tips

- I have found the greatest problem in managing an aquarium is overfeeding, which quickly fouls the environment and should be avoided unless it is part of an experiment. You should feed only enough food as can be cleaned up by the fish in seven to ten minutes, twice a day. Overweight fish are not much fun to watch because they are less active, and they are likely to die more quickly than their thinner companions.
- It is possible that your aquarium will go sour at some point. Cloudy water generally indicates the growth of anaerobic bacteria, which produce toxic compounds and gasses that pollute the water. If your students observe this phenomena, the tank must be drained and cleaned. This is nothing to be alarmed about because you now have the perfect opportunity to hypothesize and discuss with your class about what might have gone wrong to cause the growth of these harmful bacteria.
- As much as possible, students should keep their hands out of the water because of soap and detergent residues. In addition to controlling evaporation, this is one of the main reasons to use a glass cover over the aquarium.
- If more algae begins to grow than your snails can consume, it can easily be removed by using a sponge scraper. Do not use a metal scraper on a plastic aquarium since it will scratch the walls. Inexpensive sponge scrapers are made specifically for plastic tanks. If algae growth becomes a serious problem, it is probably because the tank is getting too much sunlight. To discourage algae growth, use fluorescent lighting to illuminate the tank rather than incandescent lighting.
- Be sure to place the air pump at a level higher than the water level in the tank. If you don't do this and there is a power failure, water may siphon out of the tank, possibly draining it and ruining your air pump.
- Keep a small "hospital" tank available. Sick fish should be removed from the classroom aquarium as soon as they are spotted. Like any other animal, fish are subject to various diseases. One of the most common is a disease called "ich," or *Ichthyophthirus*. Most fish diseases can be treated in a hospital tank by adding the appropriate medicine to the water. I have to admit, however, that I don't spend much time treating sick guppies. I usually destroy them when the children aren't around. If I was keeping expensive exotic fish, I would probably spend more effort on treating their diseases.
- The acidity or alkalinity of a substance, such as the water in an aquarium, is noted by its pH. A pH of 7 is neutral. Any pH below this figure indicates an acid substance, while a pH above 7 indicates an alkaline substance. Guppies and goldfish can thrive in water with a pH range of 6 to 7.5. With most community water supplies, the pH should not be a problem. Over a long period of time, however, aquarium water has a tendency to become acidic. Under-gravel filters tend to accelerate this process. If you are trying to maintain a classroom aquarium through the school year, your students should use a pH kit to periodically check the pH of the water. If the water is too acidic, add very small amounts of sodium bicarbonate over a period of several days, checking the pH several hours after each addition. This time period allows the chemical to be mixed well in the tank. If the water is too alkaline, add very small amounts of sodium biphosphate to the water using the same method.

Enrichment Activities

- One of the most fascinating types of aquariums is a saltwater aquarium, or an artificial tidepool. Most saltwater aquariums are very specialized, and you will want to find out all you can about them before trying to establish one. However, there is an easy saltwater activity you can do with your students. This simple saltwater aquarium doesn't require specialized or expensive materials, and is used to raise brine shrimp. Brine shrimp eggs can easily be obtained through pet shops or science equipment suppliers. The following list of materials and activity procedures will get you started.

  Raising Brine Shrimp

  Materials: one wide-mouth quart jar, non-iodized salt or sea salt, brine shrimp eggs, aged water, aquarium heater or equivalent heat source, aquarium thermometer, brewers yeast.

  Procedure: Fill a quart jar with aged water. Add 16 teaspoons of non-iodized salt or sea salt to the water. Stir hard enough to aerate the water and dissolve the salt. Sprinkle brine shrimp eggs into the water and place the jar where sunlight will reach it at least part of the day. The water temperature needs to be kept between 75 and 80 degrees Fahrenheit. During the night, and during the day if your classroom doesn't get much sunlight, you will have to use an aquarium heater or an incandescent light bulb in a desk lamp as a heat source.

  Eggs normally begin to hatch within 24 to 48 hours. Stirring the eggs occasionally will ensure proper aeration. The eggs will hatch into a nauplius stage of development and will go through successive molts before becoming an adult. Most crustaceans go through this process.

  Feed the shrimp once or twice a week with a pinch of dried brewers yeast. Do *not* overfeed. Also, do not allow the brine shrimp to become too numerous. Overpopulation will result in undersized shrimp, depleted oxygen supplies, and cloudy water from the accumulation of too many waste products. I generally keep 20 to 30 adult shrimp in one quart of water.
- A tropical fish aquarium can be a focal point of interest in your classroom. Tropical fish often have beautiful colors and unusual appearances. Do as much research as you can before starting a tropical fish project.
- Students can do research on aquaculture or mariculture. As the world's human population continues to grow, the oceans will become an even more important source of food.
- Students can graph the numerical data from their observations (temperature, population, pH, etc.). They can also calculate the average values for this data.
- Several of your students may express an interest in establishing aquariums at home. Invite a parent or hobbyist to help these students get started.
- Visit a local pond, lake, or stream, and use a plastic bucket to collect mud and what-have-you from the bottom of the collecting site. Take this material back to your classroom and place the muck in a five-gallon aquarium that is about half full of aged water. This is known in my classroom as the "Yuck-what-is-it?" aquarium. You don't need any filter system or special equipment for this aquarium, and you should place it where it receives direct sunlight at least part of the day.

  This aquarium will not look very pleasant and it may even have an unpleasant odor, but you will have a steady supply of paramecia, amoebas, and other protists for your classroom. Your students won't have to rely on pictures, but will have the real thing to study under the microscrope.

Making a Freshwater Aquarium

1. Follow the directions of your teacher in building your classroom aquarium. Label the components of your aquarium in the diagram below.

2. Each day, carefully record your aquarium observations in your journal. Count any changes in the fish and snail populations. How much have the plants grown (in inches, centimeters, or millimeters)? What is the condition of the water? Is it clear? Does it have an odor? Is there a lot of algae growth? Where is it growing? These are just some of the questions you can answer when you observe the aquarium every day. As you note the changes taking place in the aquarium, hypothesize about the reasons why these changes are occurring.

3. List some additional experiments you would like to do with your aquarium ecosystem. (For example, you could alter the aquarium's balance by adding more fish to see what would happen.)

____

____

____

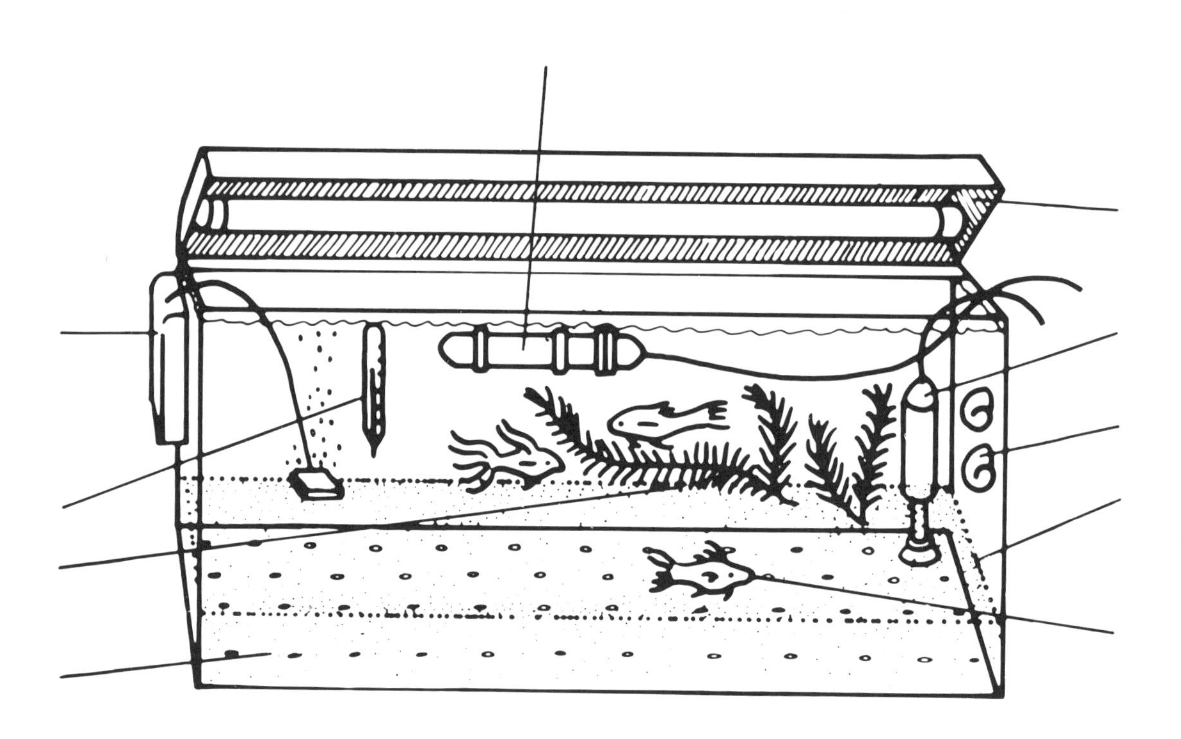

Adaptation in Aquatic Environments
Observing how animals breathe and move underwater

Group Size
2 students

Time Required
1 class period

Materials
- 1 large bucket of clear aged water (freshwater or saltwater, depending on your aquatic animals)
- An assortment of aquatic animals (guppies, goldfish, brine shrimp, water snails, tadpoles, minnows, etc.)
- 2 small bottles of food coloring (blue, green, or red work best)
- 100 ml solution of diluted food coloring (1 part food coloring to 8 parts water). Use the water from your aquatic animals' environment.

For each student group:
- 1 eyedropper
- 4 dropperfuls of the diluted food coloring solution in a paper cup
- 1 white-bottomed container (a half-gallon milk carton cut in half lengthwise will make 2 containers)
- copy of Handout 1, "Adaptations for Underwater Life," for each student
- each student should have a journal for recording observations
- 1 hand lens (optional)

Key Term
adaptation

Instructional Goal
- To identify adaptations that enable organisms to live in their particular environments.

Student Objectives
Students will:
- Observe how various aquatic animals have adapted to life underwater by experimenting with diluted food coloring to watch the currents these animals produce.
- Define adaptation.
- Identify specific adaptations that animals make in order to survive and reproduce.

Prerequisite Knowledge
Students should understand the material from Lesson 1, particularly the concept of an ecosystem. Students should also know what the term *adaptation* means.

Advance Preparation Time
About 1 hour
- Review the "Background Information" and "Lesson Procedure."
- Gather materials and aquatic animals.
- Duplicate the student handouts.
- Make sure to allow your aquatic animals time to adjust to the new water temperature. Use a plastic bag and water from their old tank to float the animals on top of the new water for about one-half hour before you release them. Also, be sure to age (dechlorinate) any tap water you use by allowing it to stand for at least 24 hours before placing animals in it.

Teacher Tips
- Encourage student interaction and comparisons, but remind students not to carry their containers around the room to show others. This is how spills and other accidents take place.

Background Information

When aquatic animals move underwater, they create a variety of currents. These currents are caused by the swimming, feeding, and breathing activities of the animals. These currents are usually difficult to observe because of the lack of color contrast between the currents and the surrounding water.

You can solve this "invisible current" problem by using drops of diluted food coloring in the water, which safely passes through animals that actively pump water through their bodies for respiration. This technique provides enough color contrast to make the currents visible to students.

In this activity, each student group places an aquatic animal (freshwater or saltwater) into a container filled with clear water and, with the aid of an eyedropper and diluted food coloring, observes the breathing, swimming, and feeding currents caused by the animal. Students can also investigate the responses of their aquatic animals to other animals, plants, and non-living objects taken from the animals' environment. These investigations allow students to observe the adaptations that allow aquatic animals to live underwater. For the purposes of this lesson, an adaptation is defined as any special feature of an organism that enables it to survive and reproduce in a particular area. The activity closes with students considering what adaptations they would need to live underwater.

Lesson Procedure

Step 1:
Collect a variety of aquatic animals a day or two before the activity, such as the guppies, goldfish, or brine shrimp from Lesson 1; or tadpoles, water snails, minnows, etc., from a local lake, stream, or pond. Also collect some plants and non-living objects from the animals' environments. If possible, collect enough aquatic animals to allow your students to investigate more than one kind of animal. To make sure your animals are fresh and lively, keep them in uncovered containers with lots of aged water, or the natural water they came from. If you use jars, be sure to use wide-mouth ones.

Step 2:
Dilute the food coloring ahead of time by mixing one part food coloring with eight parts of water from the collection site. Make about 100 ml of the food coloring solution.

Step 3:
On the day of the activity, explain the term adaptation, and tell your students that they will be investigating the adaptations shown by several types of aquatic creatures that enable them to live underwater. Explain that many aquatic animals create a variety of underwater currents when they breathe, feed, and swim. These currents are usually difficult to observe, but by using an eyedropper and diluted food coloring, students can easily see these currents. Distribute the student handouts, and divide your class into groups of two.

Step 4:
Demonstrate the proper procedure for this activity by placing an active aquatic animal in a container filled with clear water. Fill an eyedropper with the diluted food coloring solution. Place the tip of the eyedropper underwater and close to the mouth of the animal being investigated, and release *one drop* of the diluted food coloring. Only one carefully placed drop at a time is needed. Have students observe what happens, then repeat the procedure by placing a drop of diluted food coloring near the animal's tail, fins, etc.

Step 5:
Have students get their materials and let each student group choose an aquatic animal to investigate. Students should use the eyedropper technique to search for currents caused by their animals' breathing, feeding, and swimming movements. Challenge students to discover as much as they can about the currents their animals create, especially breathing currents. Encourage students to apply the food coloring near all parts of the animal, not just the head.

Step 6:
As students are working, refill their containers with clear water when needed. (Be careful not to lose the aquatic animals during the refilling process.)

After about 30 minutes, have your students investigate their animals' reactions to other organisms and non-living objects taken from the animals' environments. As time allows, encourage students to investigate several other aquatic animals. When students have completed their experiments, return all animals to their original tank or natural environment.

Step 7:
Have students discuss their findings. I usually start the class discussion by asking the following questions.

- Which animals seem to breathe water by pumping it in and out of their bodies? Why do they do this?
- Do any of the animals take in or expel water from areas other then their mouths and gills? What areas?
- How do the animals move in the water?
- How does movement help an animal survive? (Helps it obtain food, oxygen, warmth, protection, etc.) The special features of an organism (such as gills for breathing or a powerful tail for fast swimming) that improve its chances for survival and reproduction are called *adaptations*.
- What adaptations would you need to live underwater (not including mechanical devices such as scuba equipment)?

To check for understanding, I ask students for written reports of their findings (or I collect and read their journals). I also ask students to list at least three of the animals used in the activity and describe one adaptation that each animal has made that is important for its survival. In addition, I have students write a definition of adaptation in their own words.

Enrichment Activities
Students can:

- ☐ Discover the adaptations of their pets, such as the cat's sharp claws that enable it to be a good hunter and to protect itself, or the dog's claws that seem to be adapted for digging.
- ☐ Research the adaptations of one (or more) wild or domestic animal that have enhanced its ability to survive and reproduce. Every animal has its own unique adaptations.
- ☐ Although this lesson deals with adaptation in animals, students can also research adaptation in plants.

Adaptations for Underwater Life

1. Choose an aquatic animal that you would like to study, and place it in a water-filled container. Your group should also have an eyedropper and some diluted food coloring solution.

2. Using the eyedropper, place just *one* drop of diluted food coloring near the head of the animal you are examining (see Figure 1 below). Watch the movement of the colored water, which indicates the currents caused by the animal. Repeat the procedure by placing a drop of diluted food coloring near a different part of your animal. Observe what happens, and write your observations in your journal for each experiment. If you like, draw pictures to show the movement of the colored water.

3. When the water in your container becomes colored and it is hard to see the currents, ask your teacher to refill your container with clear water.

4. Add another animal, a plant, or a non-living object from your animal's environment to see how your animal reacts. Be sure to write your observations in your journal.

5. Experiment with another kind of aquatic animal to see what currents it produces. Record your observations.

6. Using what you've learned, what adaptations would you need to live underwater?

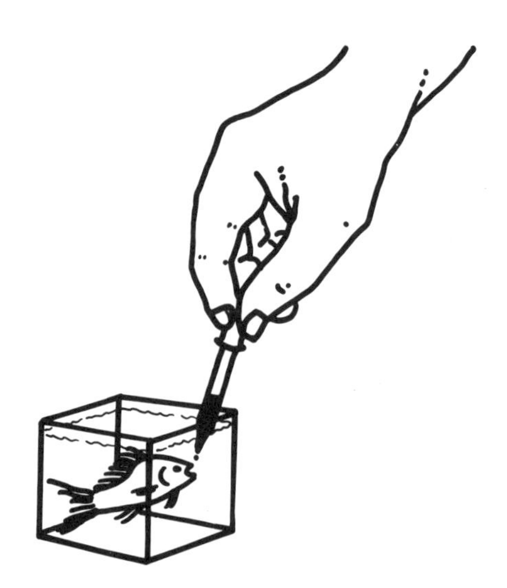

Figure 1. Place one drop of diluted food coloring near your aquatic animal, then see what happens to the colored water.

Lesson 3

Effects of Pollution on an Ecosystem

Measuring the effects of pollution on brine shrimp populations

Group Size
4 students

Time Required
1 class period

Materials
- brine shrimp eggs
- several quart-size wide-mouth glass jars (one for each pollutant, plus a control jar)
- saltwater for each quart jar (16 teaspoons of non-iodized salt or sea salt to 1 quart of aged tap water)
- samples (about 1/2 cup each) of common pollutants, such as oil (use salad oil), vinegar (to simulate acid rain), various soaps and detergents, fertilizers, insecticides and herbicides (use with caution—see notes under "Teacher Tips")
- aquarium heater for each jar
- aquarium thermometer for each jar
- aquarium net
- copy of Handout 1, "Observing the Effects of Pollution" for each student
- each student should have a journal for recording observations

Key Terms

population	pollution
ecosystem	pollutant

Instructional Goal
- To demonstrate the effects of various kinds of pollution on an ecosystem.

Student Objectives
Students will:
- Make predictions (hypotheses) about the effects of pollutants on brine shrimp populations, and on the hatching success of brine shrimp eggs in polluted water.
- Record the data obtained from observing the effects of pollution.
- Compare the results of the experiment with their original hypotheses, and suggest additional experiments.

Prerequisite Knowledge
Students should understand the key terms, as well as the concept of an ecosystem (from Lesson 1).

Advance Preparation Time
About 2 hours
- Review the "Background Information" and "Lesson Procedure."
- Gather materials.
- Duplicate the student handouts.
- Two or three days before the activity, prepare a supply of saltwater (the amount will depend on how many quart jars you use). Be sure to use aged tap water that has been allowed to stand for at least 24 hours.
- When you have prepared the saltwater and poured it into the individual glass jars, add a pinch of brine shrimp eggs to each jar. Add an aquarium heater and thermometer to each jar, and keep the water temperature between 75 and 80 degrees Fahrenheit. (See Lesson 1, "Enrichment Activities," on page 127 for more information on raising brine shrimp.) The eggs will begin to hatch within 24 to 48 hours. Each jar should have 20 to 30 brine shrimp.

Teacher Tips

- You will need to start preparing for this lesson at least two days in advance so that a viable population of brine shrimp will be ready for your students to use. See the steps under "Advance Preparation Time."
- *Caution:* If you decide to use an insecticide and/or herbicide as part of these investigations, warn your students that these chemicals are poisonous and should be handled with great care. Use only small amounts of low-toxicity chemicals, and supply students with lab aprons, rubber gloves, and goggles. Students should also wash their hands after performing the experiment. Do not dispose of these chemicals down the sink. Instead, read the label on the container and check with your district maintanence department as to the proper disposal of these substances.
- I prefer to work with students in groups of four for this activity. Larger groups tend to stray from the task, and smaller groups don't seem to interact enough.
- Some of your more inquisitive students will want to see what happens to the brine shrimp if they use greater or lesser amounts or the pollutants, or different pollutants. If time permits, allow them to conduct additional investigations. However, students should first complete the experiment as outlined on the handout so they can report their findings to the class.

Background Information

Environmental pollution is one of the most pressing problems of our times. "Better living through chemistry" may be desirable and beneficial in many cases, but we are just beginning to see and understand the effects of pollution on human health, as well as on the health of all other life on Earth. This activity is designed to graphically illustrate for your students the consequences of releasing dangerous substances into an aquatic environment.

One consequence of this experiment is the death of many brine shrimp, which you may need to prepare your students for. Explain that animals are often used in experiments that lead to the discovery of new life-saving vaccines or surgical techniques. Good examples are the development of the polio vaccine and the surgical techniques used in heart transplants.

Lesson Procedure

Step 1:
Follow the steps under "Advance Preparation Time" on page 133 to establish a brine shrimp population. This should be done two or three days in advance to allow time for the brine shrimp eggs to hatch. Be sure to have enough jars and pollutants so that each group has one to study. Also, keep a control jar of brine shrimp that will not be contaminated.

Step 2:
Distribute the student handouts, divide your class into groups of four, and assign a pollutant to each group. One member from each group can then pick up a brine shrimp jar, the pollutant sample, and any necessary safety equipment (for working with insecticides or herbicides). Have students follow the instructions on their handouts, and monitor their progress as necessary. Students

should predict what will happen to the brine shrimp when the pollutant is added to their jars, write their predictions in their journals, observe the results of the experiment, and record the results.

Step 3:
When students have completed this part of the experiment, have them use the aquarium net to remove the brine shrimp from their jars and dispose of them. Then distribute a pinch of brine shrimp eggs to each group. Students will add these eggs to their already polluted water to observe the effects of pollution on the hatching rate of the eggs. Allow 24 to 48 hours for the eggs to hatch, then have students count the number of hatchlings and record their observations.

Step 4:
After students have recorded their results of the hatching rate for brine shrimp eggs in polluted water, have students report their findings to the class. Make a chart on the board or on butcher paper to show each pollutant, its measurable effect on brine shrimp hatchlings, and the number of eggs that hatched in the polluted water (see Figure 1 below).

Step 5:
I ask the following questions during the class discussion.

- What happened to the brine shrimp in each polluted water sample?
- Why do you think the brine shrimp reacted the way they did?
- Was there a different reaction in each of the polluted water samples?
- Which pollutant had the most adverse effect on the brine shrimp? Which pollutant had the least adverse effect?
- How does the brine shrimp population in each polluted water sample compare to the population in the control jar? Why do you think it is important to have a control when conducting scientific experiments?
- What effects do you think these pollutants have when released into the natural environment? What variables can affect the magnitude of the pollutants' damage?
- If an oil spill were to occur close to shore, what do you think might happen to the sea life in that area?
- What can be done to prevent oil spills? Once an oil spill has taken place, how can it be cleaned up?
- Can you design an experiment to illustrate your clean-up method?

Pollutant	Effect on Hatchlings	Number of Eggs that Hatched in Polluted Water
Vinegar		
Oil		
Laundry Soap		
Fertilizer		
Insecticide		
Control Jar		

Figure 1. Sample chart for comparing the effects of various pollutants on brine shrimp.

To check for understanding, I collect and read the student journals. Did they accurately record the data obtained from their observations? Are their hypotheses borne out by their observations? If not, it doesn't necessarily mean the student made a mistake, but that the experiment had different results than expected. If time allows, the hypothesis can be rewritten and the experiment performed again.

Enrichment Activities

Students can:

- ☐ Read about other types of pollution (for example, air pollution, sewage and garbage disposal, nuclear waste) and write a report about how these pollutants affect the environment.
- ☐ Find out ways they can contribute to the reduction of pollution, such as recycling, proper disposal of harmful household chemicals, writing letters to government officials, and organizing neighborhood clean-ups.
- ☐ Assemble current articles from newspapers and magazines about various kinds of pollution to create a classroom bulletin board.

Observing the Effects of Pollution

1. Your group should obtain from your teacher a jar filled with saltwater that contains 20 to 30 brine shrimp hatchlings, and a small amount of your assigned pollutant. Write a prediction (hypothesis) in your journal about what will happen when the pollutant is added to the brine shrimp environment. Then place 1/2 teaspoon of your pollutant into the brine shrimp jar. Carefully observe what happens and record your observations in your journal. The brine shrimp will react immediately with some pollutants, while a longer period of time might be required for other pollutants. Watch the clock so you can record how much time it takes until you observe any reactions.

2. After you have completed your experiment with the brine shrimp hatchlings, use an aquarium net to remove the brine shrimp from your jar and dispose of them. Then place a small pinch of brine shrimp eggs in your jar. Predict how many eggs will hatch in your polluted water, and record your hypothesis in your journal. Wait 24 to 48 hours to see how many eggs hatched. Record in your journal the effect of your pollutant on the ability of the eggs to hatch. Compare your results with your original hypothesis.

3. Report your findings to the class. What other experiments would you like to do with different amounts and types of pollutants?

____

____

____

____

Lesson 4

Interdependence

Discovering what makes a balanced ecosystem

Group Size
Individual students

Time Required
1 class period initially, then daily observation for about 3 weeks

Materials
- 3 one-quart mason or canning jars with lids
- aged water to fill the jars
- 2 sprigs of *Elodea* or *Anacharis* about 4 to 5 inches long
- several aquarium animals—guppies and/or water snails
- 3 aquarium thermometers
- copy of Handout 1, "Interdependence in the Aquarium," for each student
- each student should have a journal for recording observations

Key Terms
interdependence
ecosystem

Instructional Goal
- To demonstrate that plants and animals are interdependent in a balanced ecosystem.

Student Objectives
Students will:
- Predict what will happen in three experimental aquariums, only one of which is balanced.
- Record evidence of interdependence in the aquariums and record their observations in their journals.
- Define interdependence in their own words.
- Hypothesize about the effects of heat and light on the aquariums.

Prerequisite Knowledge
Students should understand the key terms as well as the concepts about ecosystems learned in the previous three lessons.

Advance Preparation Time
About 1 hour
- Review the "Background Information" and "Lesson Procedure."
- Gather materials. Obtain the aquatic plants and animals from your mother tank (see Lesson 1, page 122).
- Duplicate the student handouts.
- One or two days before the activity, prepare about a gallon of aged water. Allow it to stand for at least 24 hours before placing animals in it.

Teacher Tips
- I recommend that you wait until your classroom aquarium is established before doing this activity. The mother aquarium can then supply you with the healthy plants and animals needed for the investigation.
- The animal(s) isolated in the airtight "animals-only" aquarium (see Step 1 in the Lesson Procedure, page 139) will probably not survive very long. I don't think it is necessary to wait until the animal dies for students to learn the concept of interdependence. I usually move the animal back to the mother tank as soon as students see signs of distress and can predict its demise if left in the jar.

- This experiment will take several weeks to complete, and you should provide time each day for students to record their observations in their journals.
- Students need to be reminded about what constitutes good observation. Otherwise, some students may give only a cursory glance to each jar and record that nothing has happened.

Background Information

This lesson explores interdependence within the aquatic biome. Students learned in Lesson 1 how to set up and maintain an aquarium so that it is balanced and can successfully support a population of plants and animals. In this activity, students learn that plants and animals in aquariums are interdependent, and need each other for their continued survival.

Plants are dependent on animals to provide nutrients (through waste products and decomposition) and, to some extent, the carbon dioxide they need for photosynthesis. Animals need plants to supply some or all of their food, as well as the oxygen necessary for respiration and metabolism. In fact, most of the oxygen used by the world's animals is actually produced by plants that live in the oceans. The three aquariums in this lesson demonstrate interdependence on a small scale, but students can easily extend this concept beyond the aquariums to include the earth's biosphere as a whole.

Lesson Procedure

Step 1:
Fill three quart-size canning jars with aged water to within an inch of their tops. Place two or three guppies, or one water snail, in the first jar. Place one 4- to 5-inch long sprig of *Elodea* in the second jar. Place two or three guppies, or one snail, *and* the other sprig of *Elodea* in the third jar. Put an aquarium thermometer in each jar, then screw on the lids to make the jars airtight (see Figure 1 on page 140). Put these mini-aquariums in a place that is out of direct sunlight, and where your students will be able to observe them easily. Maintain the water temperature at approximately 70 degrees Fahrenheit. Try to keep the water temperature in each jar the same so students can determine the effect of plants and animals on each other without introducing another variable.

Step 2:
Distribute the student handouts. Have students predict what will happen in each aquarium after several weeks, and they should record their hypotheses in their journals. Students should then observe the aquariums every day and record their observations. In addition to recording the measurable data, such as the daily water temperature, animal population, and plant growth, students should look for subtle changes in the aquariums. Are the fish active? Are they gasping for breath? Are the plants green and healthy? Is the water clear or cloudy? Does it have an odor?

The aquarium with the plants and animals should be reasonably balanced, resulting in healthy fish even after several weeks, while the fish that are alone in the first aquarium will quickly run out of dissolved oxygen. (I prefer to return these fish to their original tank when they show signs of distress.) The plant that is alone in the second aquarium may or may not show much evidence of change, depending on the amount of light it receives and how quickly it uses up its carbon dioxide and available nutrients.

Step 3:
After several weeks, have students share their observations in a class discussion. Ask some or all of the following questions during the discussion.

- Which aquarium was a model of a balanced ecosystem? What characteristics did it have that were lacking in the other two aquariums?
- Why were the other two aquariums not balanced?
- What do you think would happen if we placed all three aquariums in direct sunlight for one day? For one week?
- What would happen if we placed all three aquariums in a dark cabinet for one day? For one week?
- Where does the snail (or the fish) get its food?
- Where does the oxygen come from that the animal needs to survive?
- What does the plant need to make its food through photosynthesis?
- What was the average temperature in each aquarium?
- How does a balanced aquarium demonstrate interdependence?

Enrichment Activities
Students can:

- ☐ Perform the same experiment with different types of aquatic animals, such as goldfish or tadpoles.
- ☐ Introduce other variables into the interdependence experiment, such as water temperatures that are significantly higher or lower than the ideal of 70 degrees Fahrenheit, or placing the aquariums in sunlight or in a dark environment.
- ☐ Research other types of ecosystems and discover what interdependent components they need to stay in balance.
- ☐ Establish a plant terrarium and see if they can keep it in balance.

Figure 1. The three aquariums for the interdependence experiment.

Interdependence in the Aquarium

For this experiment, three aquariums are set up in airtight one-quart jars:

Aquarium 1. Water and aquatic animals (guppies or a snail)
Aquarium 2. Water and an aquatic plant (*Elodea* or *Anacharis*)
Aquarium 3. Water, aquatic animals, and a plant

The water temperature in all three aquariums will be the same—about 70 degrees Fahrenheit.

1. What do you think will happen in each aquarium after several weeks? Write your hypothesis in your journal.

2. Each day, observe the aquariums and record your observations in your journal. Be sure to write down the water temperature in each aquarium and any other measurable data you can think of, such as plant growth. Also be sure to note any other less obvious changes in each aquarium, such as the health of the fish and plants, and the condition of the water.

3. At the end of the experiment, compare your observaitons with your original hypothesis. Do your results confirm your hypothesis, or would you rewrite your hypothesis to fit what you know now?

4. Define interdependence in your own words. What examples of interdependence in other environments can you think of?

5. What additional experiments would you like to do with the three aquariums? What do you think would happen if you changed the amount of light or heat each aquarium received?

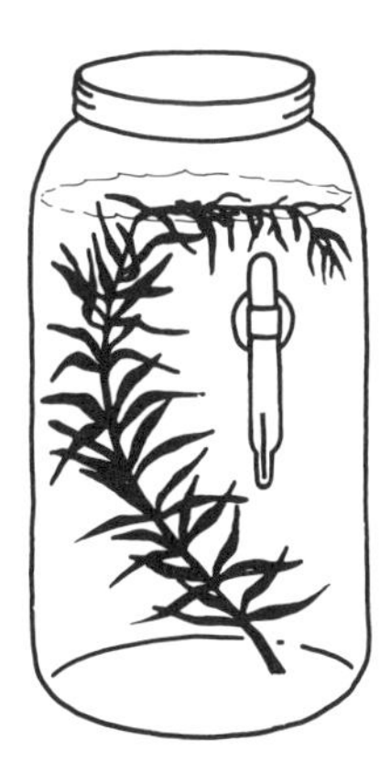

Positively Electric Activities
Basic Electricity and Circuits

Steven Oshita
William H. Crocker Middle School
Hillsborough, California

I have been a member of the science department at William H. Crocker Middle School since 1977. As part of a three-teacher team, I helped develop a laboratory-based science program that enabled our school to achieve state and national recognition as an exemplary middle school in 1984. I have also conducted workshops in science teaching at the middle school level and have co-authored an article on team-teaching in science, which was published by Far West Laboratory.

After participating in the Assessment Curriculum Evaluation Consortium as an evaluator of the Science Objective Sequence, I served on a district curriculum committee to develop a K-8 science continuum. I was also selected by GTE to participate in the 1986 Growth Initiatives for Teachers (GIFT) program in Boston, and then went on an Earthwatch expedition to Fiji to study coral communities. My goal in writing this unit, "Positively Electric Activities," was to share with other science teachers lessons that I know are successful and effective.

Lessons:
1. "BYOB: Build Your Own Battery"
2. "Simple Circuits"
3. "Measuring Electrical Energy"
4. "Electromagnetism"

Overview

HAVE YOU EVER THOUGHT about the advantages of teaching a unit on electricity? You can work with something in your classroom that doesn't require feeding, cage cleaning, or a home during vacations. It's easier to obtain than ditto paper, chalk, or iguanas. But, if you're like several other teachers I know, you might be thinking of a list of objections: you don't think you know enough about electricity to teach it, electricity is too dangerous and it scares you, you don't think your students would be interested, or you don't have the special gadgets and gizmos you need. This unit is designed to make your objections vanish by showing how you can make studying electricity fun and safe for your students.

The study of electricity is an area of physical science that is both interesting and practical. Our lives are affected every day by electricity, circuits, and electromagnetism. By studying electricity, not only will your students understand how electrical devices work, but they will also become more aware of the many ways we use electricity—from lighting lamps and heating homes to running a myriad of machines that improve the quality of our lives. Students will discover that electricity can help conserve natural resources and reduce pollution, and that it will be increasingly important to find out how to efficiently use other sources of energy, such as the sun, wind, and tides, to obtain electricity.

The lessons in this unit contain activities that are simple, safe, and guaranteed to inspire and motivate your students. You can easily find the required materials at many hardware stores or through educational supply catalogs. These activities have been successful in my classroom, and have been written to insure success in yours. The activities are hands-on and encourage student success, cooperation, and critical thinking. They are easy and fun to do, yet convey important and fundamental concepts about electricity. Impossible, you say? Not at all.

In the first lesson, students build their

own rechargeable batteries using only plastic cups, lead, cardboard blotters, and rubber bands. Using an epsom salt solution and an ordinary automobile battery charger, students can charge their batteries, connect them to flashlight bulbs, and presto! My students are always excited and full of eager anticipation when they do this activity—they can hardly wait to use their very own batteries for more experiments.

Students use their batteries to find out about electrical circuits in the second lesson. With a knife switch and a few more bulbs and sockets, students make complete, series, and parallel circuits. As part of this activity, your students can use shorthand symbols to diagram different circuits. You don't have to do research on the different types of circuits—all the information you'll need is provided in the "Background Information."

In the third lesson, students use their batteries to light bulbs, run motors, or ring bells (or whatever you may have available), and measure how much electrical energy each item requires. Students use ammeters to measure amperage and voltmeters to measure voltage in their circuits. Students also use terms they have probably seen or heard before: volts, amps, watts, and power.

The fourth lesson focuses on electromagnetism. Students wrap a wire around a nail and use their batteries to run a current through the wire. When electricity travels through the coiled wire it creates a magnetic field, and your students have made their own electromagnets. They can test the strength of their electromagnets by picking up paper clips, and then they can study the variables that determine the effects of the magnetic field.

There are, of course, many more ideas and activities that your students can do, such as testing materials for conductivity, learning about resistors and resistance, and studying induction.

This unit will give you a positive start in teaching electricity, and your students will be motivated and energetic.

Key Concepts

- ☐ Electricity is a form of energy.
- ☐ There are two kinds of electric charges: positive and negative.
- ☐ Electricity can be measured—voltage is measured in volts and amperage (current) is measured in amperes.
- ☐ Electrical power (watts) can be calculated by multiplying voltage and current.
- ☐ An electric current moves along a conductor through a complete circuit.
- ☐ Electricity can be converted to chemical energy for storage in a battery.

Skills Used in the Lessons

- observing
- comparing
- constructing
- describing
- classifying
- interpreting
- demonstrating
- predicting
- measuring
- concluding

Extensions and Sources

The lessons in this unit will help you and your students get started in exploring the study of electricity. After completing all four lessons, your students will have made a battery and studied simple circuits, measurement of electrical energy, and the properties of electromagnets. This is only the beginning; there are many more new and pertinent ideas relating to electricity that can be studied.

One logical extension to the activities in this unit is the study of conductors and insulators (or non-conductors). Students can use their batteries to test different materials in a circuit to determine which ones will conduct electricity. Another concept that is easily demonstrated using students' batteries is related to induction. Watch the movement of a wire coil carrying electricity when a bar magnet is brought near it. One end of the magnet attracts the coil and the other end repels the coil. Magnetism is related to electricity, so a unit on magnetism would be a natural extension.

I also have my students use motors and pulleys with their batteries. You can have your students make their own DC (direct current) motors and use them to pull or lift different objects. Students can then study other forms of energy, such as gravitational, potential, and kinetic. They can take this activity one step further by calculating the amount of work required to lift a small object using the motor, and then compare that with the amount of energy provided by the battery (measured with a voltmeter and ammeter). This is a basis for the study of energy conversion and heat loss due to friction, which can be used to introduce the conservation of energy principle and the first law of thermodynamics. Don't let the terminology intimidate you. Just as with electrical energy, these terms describe common events that can be understood with little effort.

Static electricity is another interesting and related topic. This can be easily demonstrated by rubbing a balloon against hair, dragging stocking feet on a carpet, or combing your hair and attracting pieces of paper to the comb. This leads to a comparison of the different types of electricity.

While this unit shares effective and successful lessons to help simplify a subject that is often thought of as complex, there are many other sources and materials available that can provide fun and simple activities. The following is a list of useful sources.

Alexander, J., et. al. *Sourcebook for the Physical Sciences.* Sacramento, CA: Harcourt Brace Jovanovich, Inc., 1967.

The Natural World I: Intermediate Science Curriculum Study. Morristown, NJ: Silver Burdett, 1981.

Peterson, Rita, and Butts, David. *Science and Society.* Westerville, OH: Charles E. Merrill, 1983.

Science '83 and Science '85. Santa Monica, CA: Enterprise for Education, Inc., 1985.

Strongin, Herb. *Science on a Shoestring.* Menlo Park, CA: Addison-Wesley, 1985.

Glossary

Alternating current (AC): A continuous back-and-forth movement of electrons in a circuit.

Ammeter: An instrument used to measure the amount of electric current in a circuit. The units are measured in amperes.

Amperage: The strength of a current in a circuit measured in amperes.

Ampere: The measure of electron flow in a circuit, expressed in coulombs per second.

Battery: A device that produces an electric current, typically through a chemical reaction.

Chemical reaction: A process in which a substance becomes another substance having different properties.

Circuit: The path of electric current flow.

Coulomb: A unit of electric charge.

Direct current (DC): The flow of electrons in one direction.

Electricity: The movement of electrons in a circuit.

Electromagnet: Metal surrounded by a coil of wire through which an electric current is passed. The metal becomes magnetized and produces a magnetic field.

Electron: A negatively charged particle in an atom.

Energy: The ability to produce motion or change.

Ion: A charged particle produced when an atom gains or loses electrons.

Kilowatt: A measure of power equivalent to 1000 watts.

Knife switch: A device with a lever that, when lowered, completes an electrical circuit.

Magnetic field: Areas or lines of force produced by magnetism.

Magnetism: A property of matter in which unlike poles attract and like poles repel.

Parallel circuit: A circuit in which two or more conductors are connected to an energy source to provide separate paths for the current.

Power: Amount of work done per unit of time. Electrical power is calculated by multiplying voltage and amperage.

Series circuit: A circuit in which the components are connected so that the same current flows through all parts of the circuit.

Volt: A unit of force pushing electrons through a circuit that produces a current of one ampere.

Watt: A unit of power calculated by multiplying volts and amperes.

Lesson 1

BYOB: Build Your Own Battery

An introduction to electricity

Group Size
2 students

Time Required
1 class period

Materials

- 1 battery charger, 6-12 volts
- 1 charging harness (a piece of wood with metal rods on each side to enable many batteries to be charged at the same time; can be made or purchased from educational suppliers)
- 1 adapter (protects against electrical shock by channeling any excess electricity to a light bulb; may be purchased from educational suppliers or made by converting a three-receptable plug from parallel to series)
- 1 100-watt light bulb
- 4 liters of magnesium sulfate (epsom salt) solution (made by dissolving 450 g of epsom salts in 4 liters of warm water)

For each group:

- 1 plastic vial and cap, about 5 cm in diameter and about 6 cm tall (purchase from an educational supplier; vials are made for this purpose)
- 2 sheets of lead, 4 cm by 30 cm
- 2 strips of lead, 4 cm by 8 cm
- 2 cardboard blotters, 5 cm by 35 cm
- 2 test leads or pieces of wire
- 1 rubberband
- 1 flashlight bulb and socket
- 2 copies of Handout 1, "Build Your Own Battery—Procedure"

Key Terms

energy	positive
electricity	negative
electron	battery

Instructional Goal

- To motivate and encourage students to learn about electricity.

Student Objectives

Students will:

- Use the materials provided to build a battery.
- Observe electrical energy being stored as chemical energy.
- Use a battery as a source of electrical energy.
- Identify uses of electricity in their daily lives.

Prerequisite Knowledge

Students should be familiar with basic lab procedures, as well as with the steps of the scientific method.

Advance Preparation Time

About 1 hour

- Review the "Background Information" and "Lesson Procedure."
- Gather materials.
- Duplicate the student handouts.
- Set up several supply areas in different parts of your classroom to make the distribution process go faster. For example, have battery supplies in one location, test leads and the epsom salt solution in another, and the charging station in another.
- Set up the battery charging system, which consists of a battery charger, a charging harness, and an adapter. See "Lesson Procedure" for complete directions.
- Mix the epsom salt solution and put it in a dispensing container that can be closed.

Teacher Tips

- Stress safety. Students must develop a healthy respect for electricity. After charging their batteries, be sure students disconnect their test leads from the charging harness first, then from the battery terminals. Test leads should not be allowed to touch together. Students should never play with wall outlets or touch exposed metal surfaces that have been connected to an electrical source. Have students wash their hands thoroughly after working with the materials used in these lessons.
- Set up your battery charging station in an open and easily monitored area. You should supervise each group's battery and connections to the charging harness. If your class is large, you may want to set up two battery charging systems to help the process go faster.
- Use a plastic soap bottle, such as a liquid dishwashing soap bottle, for dispensing the magnesium sulfate (epsom salt) solution. This makes handling easier and prevents spills. The epsom salt solution is harmless, but if it is spilled or not emptied out of the batteries at the end of the class period, the water will evaporate and leave a white powder residue. If the battery tabs are covered with this powder, it may interfere with a good contact between the tabs and the test leads.
- I find this activity works best when students are teamed up in pairs rather than working individually. This promotes cooperation and interaction, and also cuts down on the amount of materials needed. I do not advise using groups of three or more because someone is usually left out.
- I find it's helpful to demonstrate the first steps of the procedure outlined in the student handout, particularly folding the lead strips and attaching them to the lead sheets.
- Monitor and facilitate your students' progress on an individual basis. Encourage less confident students, challenge the faster ones, and check their answers and procedures for understanding. Keep students on task as much as possible, and handle disruptive problems immediately. Stress student responsibility, safety, cooperation, organization of materials, and proper handling of equipment. Students must understand your expectations and limits.
- I try not to give direct answers to student questions. Rather, I respond to questions with other questions that will lead students toward the answer. I find that it's important to encourage problem solving, making predictions, and testing hypotheses. If you don't know the answer to a question, suggest ways that students might be able to find the answer themselves.
- If you have students who have completed the activity early but do not have time to go on to the next lesson, have them use a freshly charged battery to test the light bulbs in the equipment area. Burned out light bulbs can cause frustration and should be discarded.

Background Information

In this activity, your students will build their own rechargeable storage batteries. These batteries can also be used for the other lessons in this unit or for any other activities you may want to have your students do that require batteries. I've found that this activity is a powerful motivational tool because students make something that belongs to them, and they can use their batteries to make things happen.

Depending on your science background, you may need to learn a few simple concepts about batteries and electricity before teaching this lesson. Electricity is a form of energy produced when *electrons* (negatively charged particles) travel through a wire or circuit. A complete or continuous circuit is necessary in order for electricity to flow. This type of circuit is called a *complete* or *closed circuit*. Any break or discontinuity in the circuit stops the flow of electricity, and this is called an *open circuit*. The continuous flow of electrons in a circuit is called the electric *current*.

There are two types of current commonly used, *alternating current (AC)* and *direct current (DC)*. In alternating current, the flow of electrons changes directions very rapidly. AC is the type of current that goes through the circuits in your home. In direct current, electrons in the circuit flow in only one direction. DC is the type of current produced by batteries.

The storage battery your students build in this activity produces direct current by means of chemical reactions. Charging the battery converts electrical energy into chemical energy. When the battery is used, the chemical energy is converted back into electrical energy. More specifically, the charging of the battery dissociates the magnesium sulfate (epsom salt) solution into individual ions, or charged particles. The positive magnesium ions are attracted to the negatively charged lead strip connected to the outer tab of the battery. The negative sulfate ions are attracted to the positively charged lead strip connected to the center tab of the battery.

Your students will find that the battery gets warm during the charging process and that the liquid inside bubbles. These phenomena are a result of the chemical reactions taking place. The bubbles produced by the charging process contain oxygen and hydrogen. Because hydrogen gas is potentially explosive, your students need to observe a very important safety step after their batteries have been charged. Students should disconnect their test leads from the charging system *first*, and then from the battery tabs. Disconnecting the test leads from the battery tabs first could result in a spark, which could ignite any escaping hydrogen gas. It is for this same reason that you should always be very careful not to create sparks when jump-starting an automobile.

Lesson Procedure

Preparation

Step 1:

Set up the battery charging system by positioning the battery charger, adapter, and charging harness as shown in Figure 1 on page 150. Screw a 100-watt light bulb into the socket of the adapter, and plug the battery charger into the adapter. Then connect the red and black alligator clips from the battery charger to the two metal wires or rods on the charging harness. It is important for your students to know that the black clip is negative and the red clip is positive. Label or paint one side of the charging harness for easier identification. Now plug the adapter into the wall outlet.

Step 2:

When setting up your equipment supply areas, use trays with labels so that students know what items go where at the end of the class period. I also indicate how many items are supposed to be in each tray; for example, 5 light bulbs. I group items in small numbers so I can readily see if any are missing. That way I don't have to count 30 test leads, for example. Also, designate an area for students to place their completed batteries.

You may want to include several other items in your supplies for students to use in testing their batteries, such as electric bells, buzzers, or different sizes of flashlight bulbs.

Step 3:
Stretch a wire or rope across a corner area on which to hang and organize the test leads. This way the leads hang individually and won't get tangled together like they would if they were stored in a tray or box.

Step 4:
Prepare the epsom salt solution, and store it in a dispensing container that can be closed, such as a liquid soap bottle, to prevent evaporation.

Step 5:
Cut the cardboard blotters, lead sheets, and lead strips to the appropriate sizes (or have students do this as part of the activity).

Activity
Step 1:
Discuss electricity and electrical energy with your class, including what it is, where it comes from, its uses, and required safeguards and precautions when working with electricity. Tell students what they will do in this activity, and discuss their responsibility to get materials, ask questions, follow instructions, and put everything back properly to maintain organization in the room.

Step 2:
Divide your class into pairs, distribute the student handouts, and have students get their materials. They can then start the procedure outlined on the handouts. If necessary, demonstrate how to fold the lead strips and sheets.

Monitor each group's progress, and be sure to check their batteries before they roll them. The tabs should face the same way and the lead strips should not touch anywhere.

Step 3:
When students are ready to charge their batteries, demonstrate the correct way to attach and disconnect a battery to the charging harness. The center tab should be connected to the positive (red) side of the charging harness, and the outside tab should be connected to the negative (black) side of the harness. *When disconnecting the battery, remove the test leads from the charging harness first, then from the battery tabs.*

The battery is charging when its liquid contents are bubbling. If the

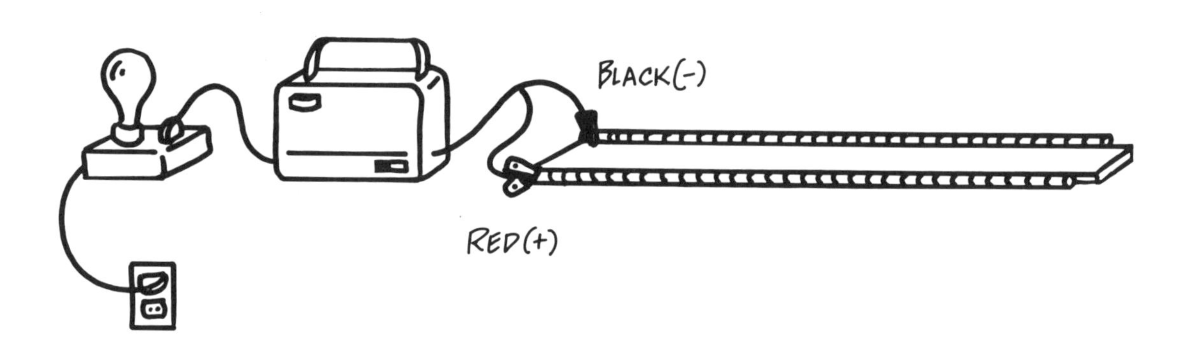

Figure 1. The battery charging system.

battery is not charging, troubleshoot the problem by making sure the charging system is plugged in, that all the clips have a good contact with the rods of the charging harness, that the center tab is connected to the positive rod, and that none of the lead pieces in the battery are touching other lead pieces.

Step 4:
After the batteries are charged, students connect their batteries to a flashlight bulb. If the bulb does not light up, troubleshoot the problem by first making sure that the light bulb is screwed into the socket all the way, that all the connections have a good contact, and that the battery has been adequately charged. When students are successful in lighting the flashlight bulbs they can try other experiments, such as ringing a buzzer or bell.

Step 5:
Signal a time to clean up about 10 minutes before the end of the class period. Be sure the magnesium sulfate solution is poured out of each battery and back into its container. Check materials and equipment to be sure everything has been returned, and that tables and desktops are clean. You'll find that after students get involved with these activities, they'll develop more efficient ways to clean up so they will have more time to experiment.

Step 6:
To review the information learned in this lesson, I have each group discuss their results and then ask for answers to the following questions.

- What is electricity? How do you know it is a form of energy?
- What are some other forms of energy?
- How is electricity stored in your battery?
- What do the terms positive and negative mean?
- What is AC? DC? Which does your battery produce? Which flows through the circuits in your house?
- Suppose you charged your battery and hooked it up to a light bulb, but the bulb doesn't light. What could be wrong? What tests could you do to find out?
- What can you use your battery for?

Enrichment Activities
Students can:
- ☐ Research the different sources, types, and uses for electricity.
- ☐ Compare different types of batteries (automobile, alkaline, nickel-cadmium, calculator, and watch) to their own storage batteries.
- ☐ Research how various electrical devices and inventions that use electricity have changed society and the world we live in.
- ☐ Compare and contrast AC and DC.
- ☐ Research the different forms of energy that electricity can be converted into, such as heat, light, potential energy, motion (or kinetic energy), and chemical energy.
- ☐ List other things they can use their batteries for.
- ☐ Develop and test hypotheses about different variables that may affect the strength of their batteries.
- ☐ Build their own DC motors, and use them in conjunction with their batteries to pull objects or perform other types of work.

Build Your Own Battery—Procedure

1. Get the following materials for your group: 1 plastic cup and lid, 2 sheets of lead (4 cm by 30 cm), 2 strips of lead (4 cm by 8 cm), 2 sheets of cardboard blotters (5 cm by 35 cm), 1 rubberband, 2 test leads, and 1 flashlight bulb and socket.

2. Place the sheets and strips of lead on a table and smooth out any wrinkles. Fold each of the short lead strips lengthwise into thirds, as in Figure 1. Now fold each strip again in half lengthwise, as in Figure 2. These pieces will be the tabs for your battery.

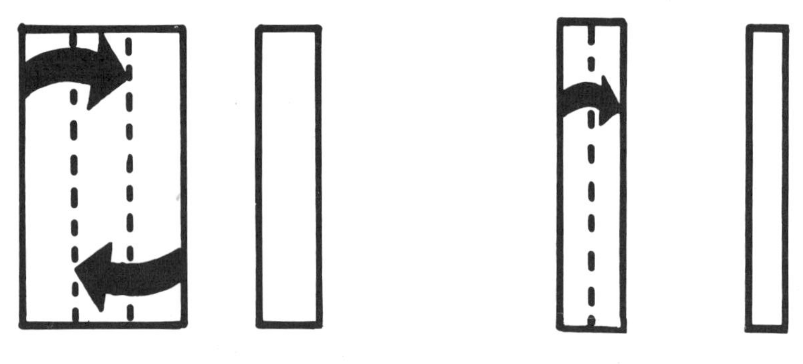

Figure 1. **Figure 2.**

3. Place one of the tabs under one end of a lead sheet, and push the lead sheet into the fold of the tab as far as you can (see Figure 3). Then fold the tab over the lead sheet once to secure it. Attach the other tab to the remaining lead sheet in the same manner.

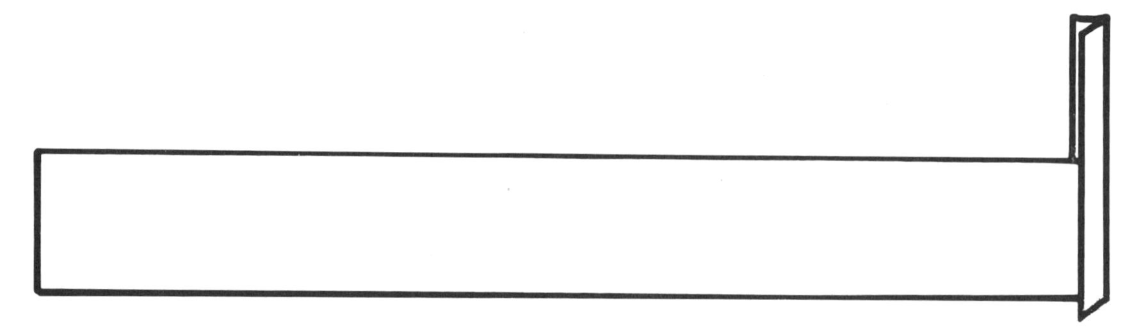

Figure 3.

4. Place one of the lead sheets on top of one of the cardboard blotters. Be sure that the lead is in the center of the cardboard. Cover the lead with the other piece of cardboard to form a lead "sandwich." Now place the other sheet of lead on top of the cardboard. Be sure that the tabs are at opposite ends of the cardboard and that they stick out in the same direction (see Figure 4).

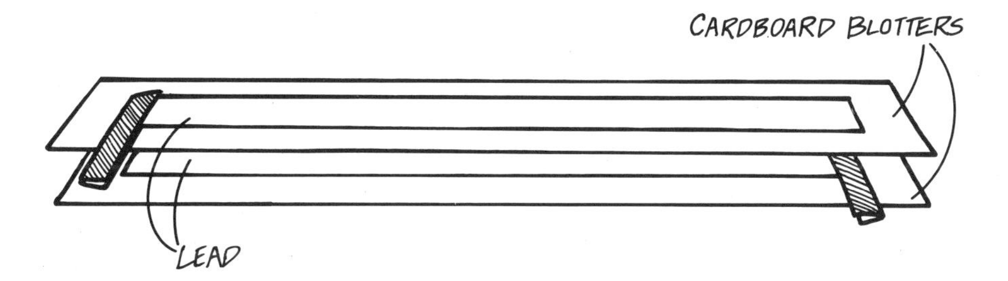

Figure 4.

5. Start at one end and roll the cardboard-and-lead sandwich into a tight roll. One tab should stick out of the center of the roll and the other tab should stick out along the edge. Wrap a rubberband tightly around the roll (see Figure 5).

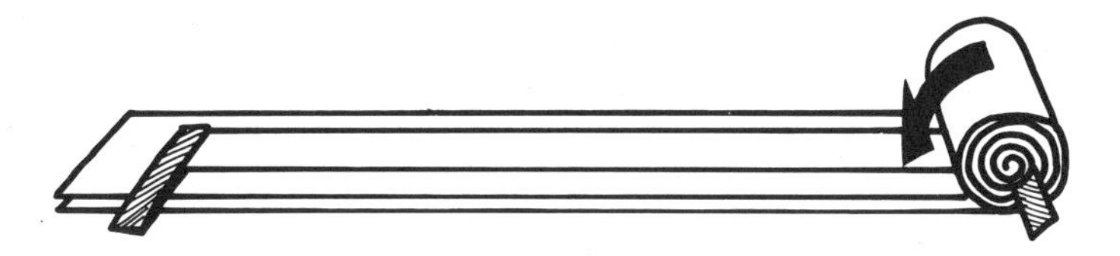

Figure 5.

6. Place the roll into your plastic cup with the tabs pointing up. Then carefully fit the tabs of your battery through the slits in the cap (see Figure 6). If the tabs fit through, you are ready to go on. If not, you may have to take the roll apart and roll it up again to reposition the tabs.

Figure 6.

7. Go to the supply area and fill your battery with magnesium sulfate (epsom salt) solution until the liquid just barely covers the roll. Then replace the cap over the roll. Your battery is now completely assembled and ready to be charged.

8. Ask your teacher to watch as you connect your battery to the charging system. Attach a test lead from the *center* tab of your battery to the *red* (positive) side of the charging harness. Connect a test lead from the *outer* tab of the battery to the *black* (negative) side of the charging harness (see Figure 7). Observe the liquid inside the battery as it charges. What do you notice? Charge your battery for 4 minutes.

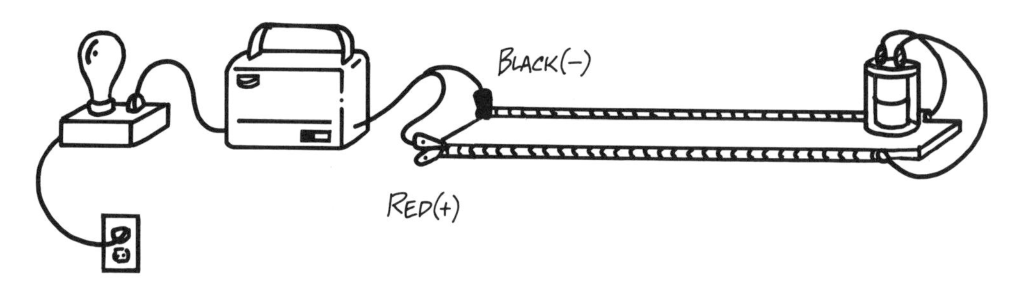

Figure 7.

9. *Important:* Disconnect the test leads from the charging harness *before* you disconnect them from your battery tabs. Do *not* let the test leads touch each other or you will shortcircuit the battery.

10. To test your battery, use the test leads to connect your battery to a flashlight bulb (see Figure 8). Does the bulb light up? If not, then something is wrong. Think about what might be the cause of the problem, then check with your teacher.

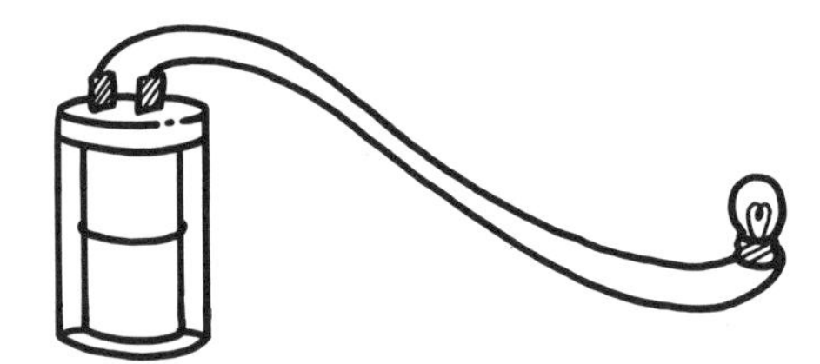

Figure 8.

11. Your battery will work better each time you charge it and use it. Continue charging your battery and testing it on a light bulb until it keeps the bulb lit for at least 1 minute.

    When it is time to clean up, pour the magnesium sulfate solution from your battery into its container. Write your group number or your names on your battery, and put it in the place designated by your teacher. Return all other equipment.

Simple Circuits
Complete, open, series, and parallel circuits

Group Size
2 students

Time Required
1 class period

Materials
- 1 battery charging system, as set up in Lesson 1
- 4 liters of magnesium sulfate (epsom salt) solution, from Lesson 1

For each group:
- 1 storage battery, made in Lesson 1
- 6 test leads
- 1 knife switch
- 3 flashlight bulbs and sockets
- 2 copies of Handout 1, "Building Simple Circuits"

Key Terms
energy
electron
current
circuit
open circuit
closed or complete circuit
series circuit
parallel circuit
short circuit

Instructional Goal
- To have students develop a basic understanding of simple electric circuits.

Student Objectives
Students will:
- Demonstrate that a complete circuit is necessary for electrons to flow.
- Identify and construct complete, series, and parallel circuits.
- Use the steps of the scientific method to describe their observations and conclusions.
- Demonstrate proper use of electrical equipment.

Prerequisite Knowledge
Students should know how to set up the required electrical equipment, and be able to perform the correct procedures for charging a battery as learned in Lesson 1.

Advance Preparation Time
About 30 minutes
- Review the "Background Information" and "Lesson Procedure."
- Gather materials and set them out at several supply stations.
- Duplicate the student handouts.
- Set up the battery charging system as in Lesson 1 (see pages 149–151 for directions).
- Mix up additional magnesium sulfate solution if you don't have enough left over from Lesson 1 (see the formula on page 147).

Teacher Tips
- Stress safety, especially when students are charging their batteries. To disconnect a battery from the charging harness, *always* disconnect the test leads from the harness *first*, then from the battery tabs. Be sure to monitor students at the battery charging station.
- Student should wash their hands thoroughly after handling the materials and equipment.
- Use more than one battery charging station if your class is large.

Background Information

Electricity travels through wires, and these wire paths are called *circuits.* Electric circuits are all around us—in the wiring of our homes, in appliances and automobiles, and in overhead and underground electrical wires. Sometimes these circuits are hidden, such as in the walls of our homes or in underground cables. For electricity to flow, these circuits must not have any gaps or breaks. This continuous electrical path is called a *complete* or *closed circuit.* A circuit that has a break in it that prevents the flow of electricity is called an *open circuit.* By opening and closing circuits we are able to control the flow of electric current, such as when we turn lights on and off by using switches. Just think of how many times a day *you* control electricity.

This lesson introduces two simple and very common types of circuits—series and parallel. In a *series circuit* all the components of the circuit are in the same line or path, therefore they share the same electric current. To remember this terminology, think of the World Series, which consists of several baseball games played one after the other. In a series circuit one component receives the electricity, uses some of it, then passes the remainder on to the next component, and so on. A problem arises, however, when one component does not pass on the electricity—it creates a break in the circuit. The result is an open circuit, and none of the components in the series will work. A common example of a series circuit is a string of old Christmas tree lights—if one bulb goes out, they all go out.

A *parallel circuit* solves this problem. In a parallel circuit each component has its own path to the energy source, and therefore gets its electric current independent of the other components. For this reason, many newer Christmas tree lights use parallel circuits instead of series circuits. If one light burns out, the other lights are unaffected and continue to work. The circuits in your house are also parallel, so that if a light bulb in your living room lamp goes out, your other electrical devices still work.

Another type of circuit that your students should know about is a *short circuit.* In a short circuit, one terminal of a battery is connected to the other terminal. The electricity that leaves the battery goes directly back into the battery without going out to any other electrical component. This has the effect of draining the battery of its energy.

As an optional but recommended part of this lesson, you may introduce to your students the diagramming notation for simple circuits. This shorthand system was developed as the need arose to explain how to construct circuits without resorting to pictures of light bulbs, batteries, wires, switches, etc. The following symbols are used as part of this system of notation.

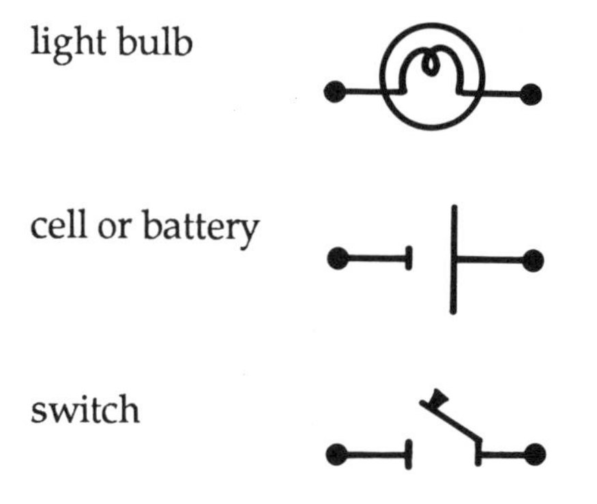

Using these symbols, a series circuit with two light bulbs, a battery, and a switch can be diagrammed as shown in Figure 1 on page 157. A parallel circuit with two light bulbs, a battery, and a switch can be diagrammed as shown in Figure 2.

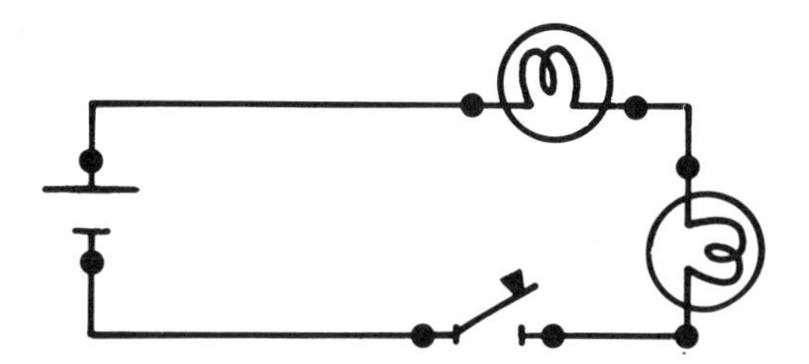

Figure 1. A diagram of a series circuit.

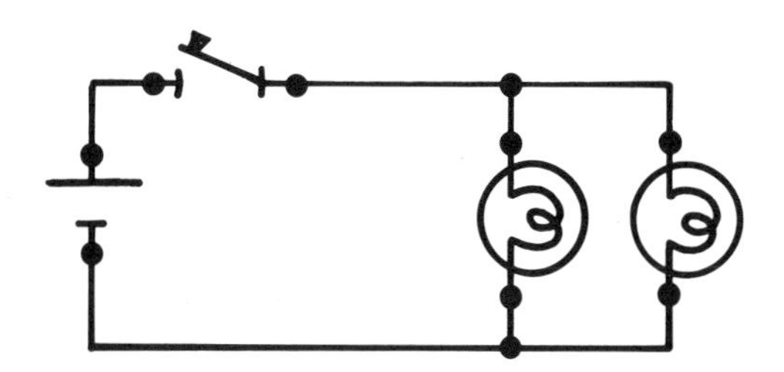

Figure 2. A diagram of a parallel circuit.

Lesson Procedure

Step 1:
Review definitions of electrical terms with your students, such as energy, electricity, electron, and circuit. Discuss various examples of circuits, where they are found, and what they are used for. Explain the difference between open and complete circuits, and between series and parallel circuits. (You might even bring in strings of Christmas tree lights in series and parallel for demonstration.) Ask students what a switch does in a circuit. Tell your students that in this activity they will use the batteries they made in Lesson 1 to build their own series and parallel circuits. Also be sure to review the battery charging system and the correct way to safely disconnect a charged battery from the charging harness.

Step 2:
Students should work in the same pairs as in Lesson 1. Distribute the student handouts, and have students get their materials. They will need to refill their batteries with magnesium sulfate solution, and then they should charge their batteries for about 4 minutes. When these preparatory steps have been completed, students are ready to go on with the procedure outlined in the student handouts.

Step 3:
If students have trouble getting their circuits to work correctly, troubleshoot the problem by making sure the battery is charged, that all the connections have a good contact, and that all the light bulbs are securely fastened in their sockets. Always have students do as much troubleshooting as they can on their own. Have students complete their circuits and write their observations and conclusions on their handouts.

Step 4 (optional):
Discuss the notation system that is used to diagram circuits. Have students draw diagrams of their circuits using the symbols shown on page 156.

Step 5:
Allow about 10 minutes at the end of the class period for clean up and return of equipment. Make sure students pour the magnesium sulfate solution out of their batteries and back into the dispensing container.

Step 6:
Schedule time at the end of the class period or at the beginning of the next class to have students discuss their results. To check for understanding, I ask my students to answer these questions:

- What is the difference between series and parallel circuits? Which type is found in your house? In school? In a flashlight? How do you know?

- What did the switch do in your experiment? When do you see switches being used every day? Were the switches in your experiment used for the same purpose?
- What is a circuit breaker? Do you know where the circuit box is located in your home? How do you use it? When would a circuit breaker be needed?
- What is a short circuit? When have you heard the term used before?
- What would happen if two batteries were set up in parallel? Predict what effect this would have on a light bulb inserted into the circuit. Draw a diagram to show this arrangement.
- What would happen if two batteries were set up in series? Predict what effect this would have on a light bulb inserted into the circuit. Draw a diagram to show this arrangement.
- Older Christmas tree lights were usually strung in series, although they are commonly strung in parallel now. If you had a string of Christmas tree lights in series and one of the bulbs burned out, how could you find which bulb (or bulbs) was the offender?

Enrichment Activities

Students can:

- ☐ Try devices other than light bulbs, such as buzzers or bells, in series and parallel circuits. Have students diagram their circuits and get your approval before constructing them.
- ☐ Try adding more light bulbs or other devices to see how this affects their batteries.
- ☐ Research the development of different types of circuits.
- ☐ Bring in circuit diagrams that come with television sets, speakers, etc., and identify as many types of circuits as possible.
- ☐ Find out how the circuits are set up in a home—are they in series or parallel or both? How about batteries in a flashlight, or the circuits in an automobile?
- ☐ Build a small-scale model home, such as a doll house, as a class project. Wiring can be run throughout the rooms, using switches and batteries to turn the lights in each room on and off. Students can then diagram the circuits in their model home.

Building Simple Circuits

An electric circuit is a pathway through which electricity travels. There are two basic types of circuits—*series* and *parallel*. In the following experiments you will find out how these two types of circuits work.

1. Get the following materials for your group: 6 test leads, 1 knife switch, 3 flashlight bulbs and sockets, and your storage battery. Refill your battery with magnesium sulfate (epsom salt) solution, then recharge your battery for 4 minutes.

2. Using your storage battery, a switch, a light bulb, and 3 test leads, set up a circuit as show in Figure 1. Make sure your battery is fully charged. Notice that the switch should be open when you set up the circuit.

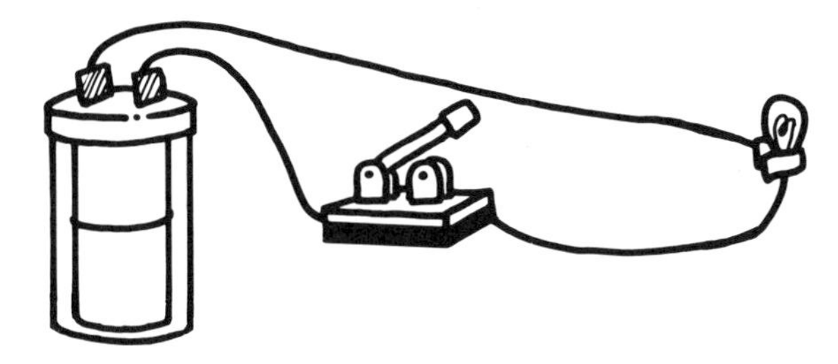

Figure 1.

Now close the switch. What do you observe? ______________________

____

If the bulb did not light up, check to make sure that your circuit is set up correctly and that all the test leads have a good contact.

3. Open the switch and disconnect one of the test leads from your battery. Then close the switch. What do you observe? Explain your observation.

____

____

4. Open the switch and connect the test lead to the battery. Then close the switch and unscrew the light bulb several turns. If the light bulb goes out, what do your know about the circuit?

____

To get electricity from the battery to the light bulb you must have a continuous path, which is called a *closed* or *complete circuit*. A break or opening in the circuit can be caused by a burned out or unscrewed light bulb, an open switch, or by poor connections with the test leads. When there is a break in the circuit, it is called an *open circuit*.

5. Recharge your battery for 3 or 4 minutes. Set up another circuit using 3 light bulbs as shown in Figure 2.

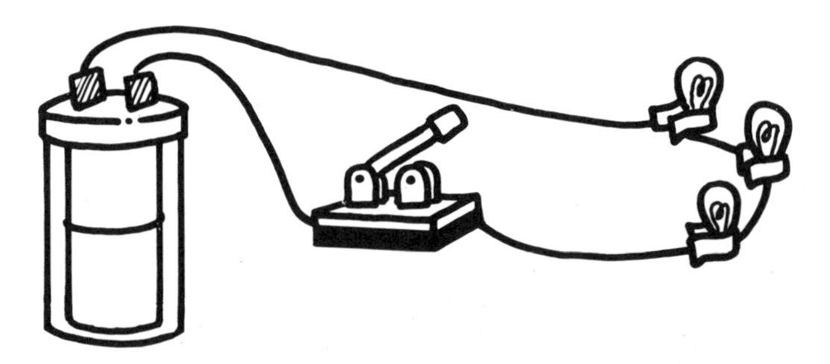

Figure 2.

Close the switch so that all the light bulbs are lit. Disconnect one of the wires in the circuit. What happens?

____

6. Reattach the wire that you disconnected. Be sure that the switch is closed and that all the light bulbs are lit. Now unscrew one of the light bulbs. What happened to the light bulb you unscrewed?

____

What happened to the other light bulbs in the circuit?

____

Suppose all the light bulbs in your circuit were lit and then one bulb suddenly burned out. What would happen to the other light bulbs?

____

The circuit you just built is called a *series circuit*. A series circuit has only one path through which electricity can flow. The next circuit you will build is called a *parallel circuit*. A parallel circuit has several paths through which electricity can flow.

7. Recharge your battery for 3 or 4 minutes. Connect the light bulb to the battery as shown in Figure 3. Notice that each light bulb has its own pathway to the battery tabs. Make sure that all 3 light bulbs are lit. Now unscrew one of the light bulbs. Describe your observations.

____

____

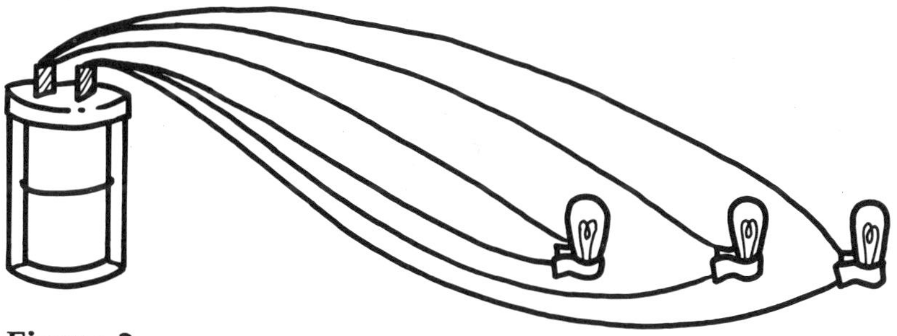

Figure 3.

8. Tighten the light bulb that was unscrewed. Now disconnect one of the wires from a light bulb socket. What happened?

____

9. You have just observed the difference between a series and a parallel circuit. Summarize what you know about these circuits.

____

____

____

____

Lesson 3

Measuring Electrical Energy

Using ammeters and voltmeters to measure electricity

Group Size
2 students

Time Required
1 class period

Materials
- 1 battery charging system, as set up in Lesson 1
- 4 liters of magnesium sulfate (epsom salt) solution, from Lesson 1

For each group:
- 1 storage battery, made in Lesson 1
- 1 ammeter
- 1 voltmeter
- 1 knife switch
- 3 flashlight bulbs and sockets
- 8 test leads
- 2 copies of Handout 1, "Measuring Electrical Energy"

Key Terms

current volts
amperes voltage
amperage watts
power

Instructional Goal
- To develop an understanding of how electricity can be measured.

Student Objectives
Students will:
- Learn the meaning of the terms amperage, voltage, power, and watts.
- Use ammeters and voltmeters to measure amperage and voltage in circuits.
- Calculate power (in watts), voltage (in volts), and amperage (in amperes).
- Demonstrate the proper use of electrical equipment.
- Demonstrate how to read the voltmeter and ammeter scales correctly.

Prerequisite Knowledge
Students should be familiar with the material in the first two lessons of this unit, including knowledge of series and parallel circuits, and the proper procedures for charging a battery.

Advance Preparation Time
About 30 minutes
- Review the "Background Information" and "Lesson Procedure."
- Gather materials and set them out at several supply stations.
- Duplicate the student handouts.
- Set up the battery charging system as in Lesson 1.
- If necessary, mix up additional magnesium sulfate solution to make 4 liters (see the formula on page 147).

Teacher Tips
- Emphasize safety. Monitor students as they charge their batteries to make sure they disconnect the test leads from the charging harness first, then the battery tabs.
- Students should wash their hands after handling the materials in this lesson.
- Caution students not to carry out any additional experiments on their own without your permission.

Background Information

Many of the terms used when studying and measuring electrical energy are common, such as volts, amps, watts, and power. We use these terms to read an electric bill, find the correct size of batteries, or buy the right size of light bulbs. Your students will also use these terms in their everyday lives. But what exactly do these terms mean?

Think of the electric current in a circuit as running water in a hose. Just as water is made up of many tiny water molecules, electric current is made up of tiny particles of electrical energy called electrons. The speed or rate at which these electrons flow in a current is expressed in *amperes*, or *amps*. An *ammeter* is a device used to measure amperes in a series circuit. Some appliances require more energy to work, so they draw electrons at a higher rate, or more amperes. For example, a hair dryer draws more energy, or amperes, per second than a light bulb.

Another unit of measure of electricity is the *volt*. A volt is defined as a unit of electromotive force. In other words, volts measure how much energy each electron has. Voltage can be compared to water pressure in our water hose analogy. A flashlight battery, which has a voltage of 1-1/2 volts, is similar to low water pressure; an automobile battery that delivers 12 volts is similar to medium water pressure; and the alternating current (AC) running through our homes, which is 110-120 volts, is similar to high water pressure. The higher the voltage, the stronger the electrical energy. To measure voltage, a *voltmeter* is set up in a parallel circuit.

The third unit of measure is the *watt*, which measures the power in an electric circuit. Power is calculated by multiplying the voltage and the current.

Power = Voltage x Current
(watts) = (volts) x (amperes)

For example, a 100-watt light bulb takes more power than a 40-watt bulb.

Another analogy can be used to tie all these measurement terms together. Think of a steam-powered locomotive as an object that requires a certain amount of energy to move. Two factors that determine how much energy (watts) it gets are the size of the logs that are fed to the fire (volts) and how fast the logs are put into the fire (amperes). Think about how changing the sizes and rates would affect the speed of the locomotive, and then how different voltages and currents would affect the locomotive's power.

To diagram circuits that include ammeters and voltmeters, use these symbols:

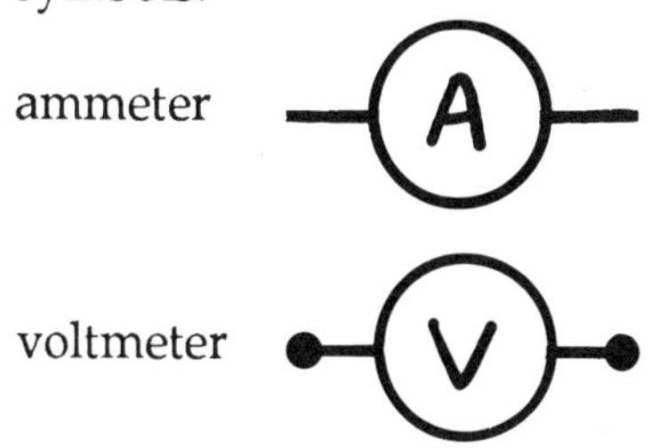

A sample diagram of a series circuit with an ammeter, two light bulbs, a switch, and a battery is shown in Figure 1.

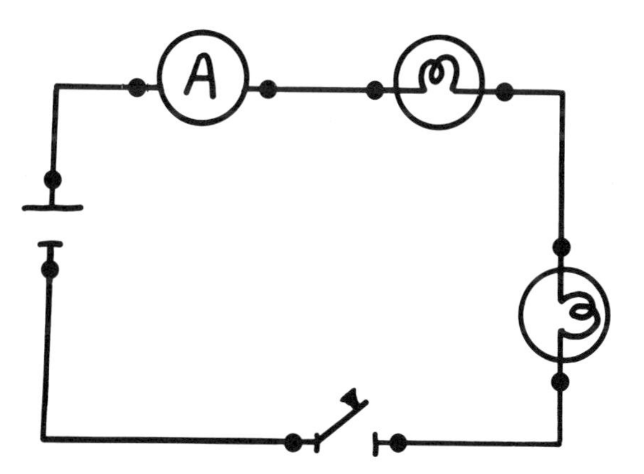

Figure 1. A diagram of a series circuit with an ammeter.

Lesson Procedure

Step 1:
Review with your students the definitions of series and parallel circuits, as well as the procedure for setting up these circuits. Explain the terms volt, amp, watt, and power, as well as how these units of measurement are related. Show your students a voltmeter and an ammeter, and demonstrate how to attach test leads to the positive (+) and negative (–) terminals of these devices. Go over how to read the measurement scales on each device, demonstrating with an actual circuit if necessary. Also be sure to review the battery charging system and the correct way to safely disconnect a charged battery from the charging harness.

Step 2:
Students should work in the same pairs as in Lesson 1, and use the batteries they made in the first lesson. Distribute the student handouts, and have students get their materials. They will need to refill their batteries with magnesium sulfate solution, and then they should charge their batteries for about 4 minutes. When these preparations have been completed, students are ready to go on with the procedure outlined in the student handout.

Step 3: (optional)
After students have completed their circuits and written their observations and conclusions on their handouts, discuss the notation used to diagram circuits that contain ammeters and voltmeters. Have your students draw diagrams of their circuits using the symbols from pages 156 and 163.

Step 4:
Allow about 10 minutes at the end of the class period for clean up and return of equipment. Make sure students pour the magnesium sulfate solution out of their batteries and back into the dispensing container.

Step 5:
Have students discuss their results and conclusions. To check for understanding, I ask my students to answer these questions:

- Why must the ammeter be connected only to a series circuit? What would happen if you connected it to a parallel circuit? To help explain your answer, remember that electrons are particles that carry energy.
- Why must the voltmeter be connected only to a parallel circuit? What would happen if you connected it to a series circuit? Explain what happens to the flow of electrons in each case.
- How is electric power calculated? Where have you heard of watts before? What is the difference between a 25-watt light bulb and a 100-watt bulb?
- What is a kilowatt? Where have you heard of kilowatts before?
- Explain how electrical devices draw current from an energy source, since the energy source does not distribute energy.

Enrichment Activities
Students can:

- ☐ Obtain an energy meter from your local utility company to determine how many watts various household appliances use. As part of this activity, have students decipher a sample electric bill.
- ☐ Compare voltages of different battery sizes. Why are batteries made in so many voltage sizes? Why aren't high voltage batteries used to run all electrical devices?
- ☐ Run a DC motor with their storage battery, and measure the voltage and current. Then try to restrict the movement of the motor by holding the shaft. How do the voltage and current measurements compare?
- ☐ Research other devices that measure electricity, and draw circuit diagrams using these devices.

Measuring Electrical Energy

In this activity you will measure the amount of electrical energy used by a component in a circuit. This is done by using a voltmeter to measure voltage and an ammeter to measure amperage.

1. Get the following materials for your group: 1 voltmeter, 1 ammeter, 3 light bulbs, 1 knife switch, 8 test leads, and your storage battery. Refill your battery with magnesium sulfate (epsom salt) solution, then recharge your battery for 4 minutes.
   While your battery is recharging, study the ammeter and voltmeter so you can read the scales on the meters. The units of measure for current are *amperes* or *amps*. The units of measure for voltage are *volts*.

2. Connect the ammeter in a series circuit as shown in Figure 1. Close the switch and read the ammeter. What is the ammeter measurement? Record this number in the first column of the chart below.

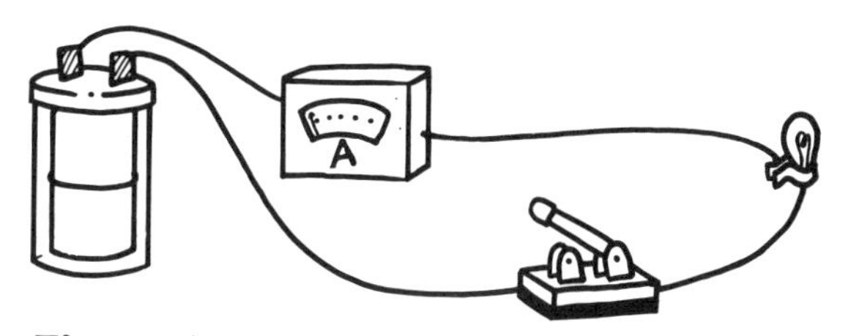

Figure 1.

Number of Bulbs	Ammeter Reading (Amps)	Voltmeter Reading (Volts)	Power Calculation (Watts)
1			
2			
3			

3. Open the switch. Then connect another light bulb in the series circuit as shown in Figure 2. Close the switch and read the ammeter. Record your data in the chart on page 1. Then open the switch.

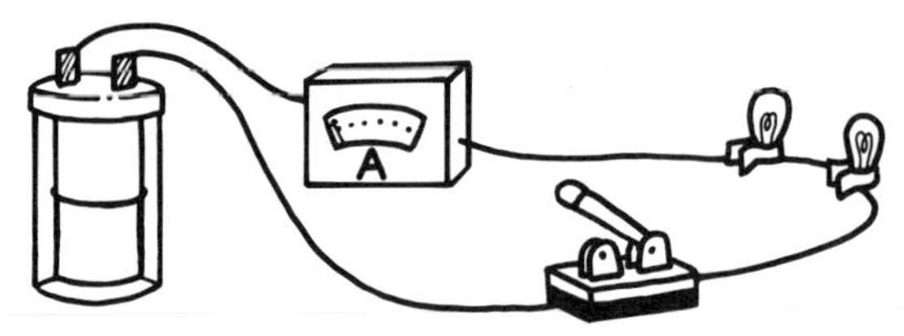

Figure 2.

4. Attach a third light bulb in your series circuit. What do you predict the ammeter reading will be?

   ____

   Close the switch and read the ammeter, then record the reading in the chart on page 1.

5. Recharge your battery again. Using what you know about current and amperes, how would you explain the results shown in the chart?

   ____

   ____

6. Set up a circuit as shown in Figure 3. Be sure the switch is open. Notice that the ammeter is set up in series and the voltmeter is set up in parallel to the light bulb. To correctly measure electrical energy, the ammeter and voltmeter must be set up only in these types of circuits.

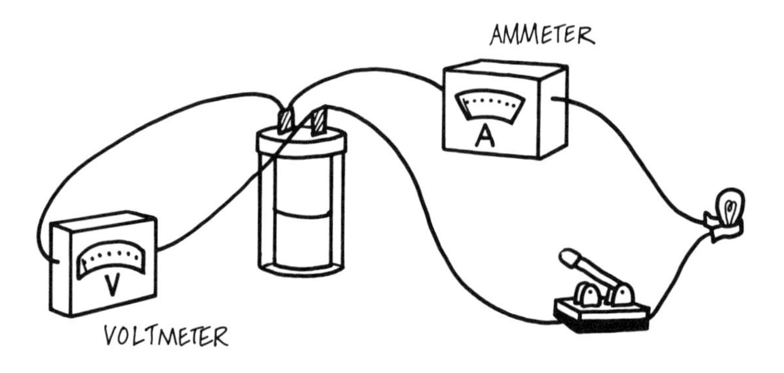

Figure 3.

With the switch in your circuit open, what does the ammeter read? What does the voltmeter read?

____

Now close the switch. Record the ammeter and voltmeter readings in the chart on page 1.

If your circuit was working correctly, you should have found that the voltmeter reading was the same whether the switch was open or closed. Using what you know about voltage and volts, explain why the voltmeter readings were the same.

____

____

7. Predict what would happen to the voltmeter readings if you added a second and then a third light bulb to the circuit.

   ____

   ____

   Test your prediction by adding the light bulbs to the circuit. Record the resulting voltmeter measurements in the chart on page 1. What can you conclude about your prediction?

   ____

8. Everything that uses electricity to operate uses *power*. Power is measured in units called *watts*. To calculate how much power is used by a device, multiply the voltage and the current. The formula is:

   Power = Voltage x Current
   (watts) = (volts) x (amperes)

   Calculate how much power one light bulb used in your circuit. Record your answer in the chart. Then calculate the amount of power used by two and three light bulbs in your circuit. Record your answers in the chart. What are your conclusions about the amount of power used in your circuit?

   ____

   ____

Lesson 4

Electromagnetism

Building and using an electromagnet

Group Size
2 students

Time Required
1 class period

Materials
- 1 battery charging system, as set up in Lesson 1
- 4 liters of magnesium sulfate (epsom salt) solution, from Lesson 1

For each group:
- 1 storage battery, made in Lesson 1
- 2 pieces of wire, one 20 cm long and one 30 cm long (I use uninsulated copper wire)
- 1 nail or screw about 5 cm long
- 1 knife switch
- 20 metal paper clips
- 1 bar magnet
- 3 test leads
- 2 copies of Handout 1, "Build Your Own Electromagnet"

Key Terms
magnet
electromagnet
poles
attract
repel

Instructional Goal
- To develop an understanding of how an electromagnet works.

Student Objectives
Students will:
- Observe the relationship between electricity and magnetism.
- Demonstrate that electricity flowing through a wire produces a magnetic field.
- Compare the strength of an electromagnet to the length of wire wrapped around it.
- Demonstrate the proper use of electrical equipment.

Prerequisite Knowledge
Students should be familiar with series and parallel circuits, magnetism, and the proper procedure for charging a battery.

Advance Preparation Time
About 30 minutes
- Review the "Background Information" and "Lesson Procedure."
- Gather materials and set them out at several supply stations.
- Duplicate the student handouts.
- Set up the battery charging system as in Lesson 1.
- If necessary, mix up additional magnesium sulfate solution to make 4 liters (see the formula on page 147).

Teacher Tips
- Emphasize safety. Monitor students as they charge their batteries to make sure they disconnect the test leads from the charging harness first, then the battery tabs.
- Students should wash their hands after handling the materials and equipment in this lesson.
- Require students to get your approval before doing additional experiments.

Background Information

Electromagnets, as the name implies, are magnets made using electric current. For review, a magnet is a metal, such as iron, cobalt, nickel, or an alloy, with two poles—one that always points to the earth's magnetic north pole and one that points to the earth's magnetic south pole. The magnet's oppositie poles create a field or force that attracts or repels other magents.

A magnetic field can also be produced by the flow of electric current in a wire. This creates a temporary magnet called an electromagnet. The stronger the force causing the flow of electrons (voltage), the stronger the magnetic field. Also, the more wire available to carry the electrons, the more area for the magnetic field to act. Therefore, two variables that determine the strength of an electromagnet are voltage and the length of wire.

In this lesson, your students will demonstrate electromagnetism by wrapping a wire around a nail or screw and running a current through the wire. The result is a temporary electromagnet that your students can experiment with and study. Your students will also measure the strength of their electromagnets by determining the number of paper clips they can attract. Then, using the steps of the scientific method, students can experiment to find out what happens if they wrap the nail with more wire, or if they increase the strength of the battery by varying the charging time. As additional experiments, students can also study the effects of attaching two or three batteries to the electromagnet, or adding other electrical components to the circuit.

This lesson is simple and conveys an important conceptual foundation that encourages further study and testing, problem solving, creative thinking, and cooperation. All that is needed is your guidance, enthusiasm, and motivation.

Lesson Procedure

Step 1:

Review the properties of magnets and magnetism with your students. If they have never studied magnetism, you may want to do an introductory lesson on magnets. Explain the term electromagnet. Review the correct procedures for charging a battery.

Step 2:

Students should work in the same pairs as in Lesson 1, and use the batteries they made in the first lesson. Distribute the student handouts, and have students get their materials. They will need to refill their batteries with magnesium sulfate solution, and they should charge their batteries for about 4 minutes. When these preparations have been completed, students are ready to go on with the procedure outlined in the student handout. Monitor their work, and when possible try to have students find the answers to their own questions through experimentation and by using the steps of the scientific method.

Step 3:

The last question on the handout asks students to design additional experiments to determine the variables that affect the strength of their electromagnets. Be sure students get your approval for their experiments before they start.

Step 4:

Allow about 10 minutes at the end of the class period for clean up and return of equipment. Make sure students pour the magnesium sulfate solution out of their batteries and back into the dispensing container.

Step 5:
Have students discuss their results and conclusions, as well as the additional experiments they conducted. To check for understanding, I ask my students to answer these questions:

- Did you find poles on your electromagnets that were similar to the poles on a bar magnet?
- How are electromagnets different from permanent (or bar) magnets?
- Do electromagnets attract and repel each other in the same way that permanent magnets do? Why or why not?
- Do electromagnets attract or repel permanent magnets in the same way that two permanent magnets attract or repel each other?
- Could you change the strength of your electromagnet? How?
- Did you find that your paper clips became magnetized? How did this happen?
- Can you think of some uses for electromagnets?

Enrichment Activities
Students can:

- ☐ Experiment to see how two or more electromagnets affect each other.
- ☐ Try to magnetize different items, and draw conclusions about the properties of magnetism.
- ☐ Insert light bulbs in their circuits to test what effect the bulbs have on the strength of their electromagnets.
- ☐ Find out how two batteries connected in series and in parallel would affect the strength of their electromagnets.
- ☐ Research how a magnetic field can produce magnetism (induction).
- ☐ Research the invention of electromagnets and some of their uses.
- ☐ Find out why electromagnets are used in motors, bells, buzzers, and electricity measuring devices. How do these electromagnets work?
- ☐ Find out the effect of an electromagnet on a magnetic compass.
- ☐ Design an experiment using electromagnetism that could determine if a current is running through a wire.

Build Your Own Electromagnet

1. Get the following materials for your group: 2 pieces of wire (one 20 cm long and one 30 cm long), 1 nail or screw, 20 paper clips, 1 switch, 3 test leads, 1 bar magnet, and your storage battery. Make sure your two pieces of wire have exposed ends; if your wire has plastic insulation, use scissors to remove it from the ends. Refill your battery with magnesium sulfate (epsom salt) solution, then recharge your battery for 4 minutes.

2. While your battery is recharging, take the longer wire and, starting at the middle of the wire, wrap it around the nail *exactly* 25 times (see Figure 1).

Figure 1.

3. When your battery is charged, connect one end of the wire that you wrapped around the nail to the positive (center) battery terminal. Connect the other end of the wire to your switch (see Figure 2). Be sure the switch is open.

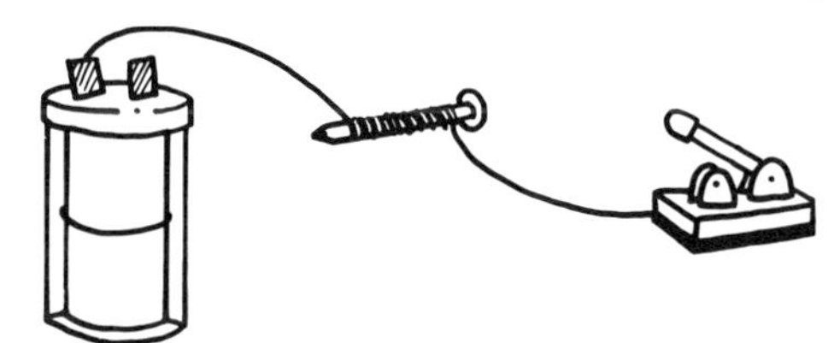

Figure 2.

4. Connect the shorter wire so that one end is attached to the negative (outer) battery terminal, and the other end of the wire is attached to the switch (see Figure 3).

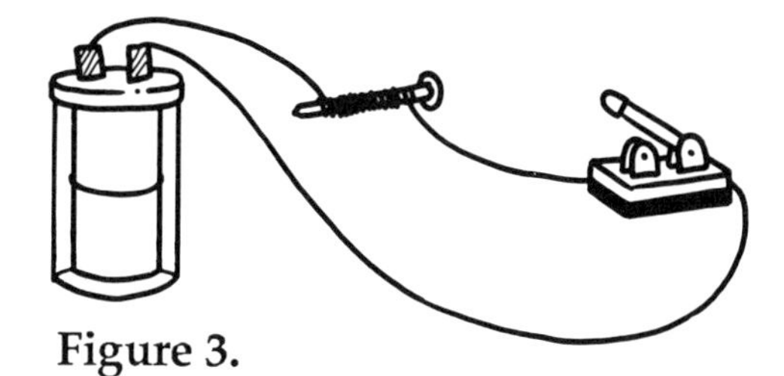

Figure 3.

5. You have just made an *electromagnet*. Hold the nail by its head and keep the switch open. Using the sharp end of the nail, try to pick up paper clips one at a time. How many paper clips can your electromagnet pick up?

   ____

   Close the switch. Now how many paper clips can your electromagnet pick up?

   ____

   From this experiment, what do you think an electromagnet needs in order to exert a magnetic force?

   ____

6. Recharge your battery for 3 or 4 minutes. While it is recharging, unwind some of the wire around the nail so that there are *exactly* 15 turns. Open the switch in your circuit. When your battery is charged, attach it to the electromagnet as in Figure 3 on page 1.

7. Hold the nail by its head and close the switch. How many paper clips can your electromagnet pick up this time?

   ____

   What can you conclude about one factor that might affect the strength of your electromagnet?

   ____

   Predict what would happen if you increased the number of wire turns around the nail to more than 25.

   ____

8. Recharge your battery and wind more turns of wire around the nail to test your prediction. How many turns did you use? How many paper clips did your electromagnet pick up?

   ____

   What can you conclude about your prediction?

   ____

9. Recharge your battery for 3 or 4 minutes, then reattach your battery to the electromagnet as shown in Figure 3. With the switch open, hold the nail by its head. Close the switch and bring the sharp end of the nail near your bar magnet. What happens?

____

10. Open the switch. This time hold the nail by the sharp end. Close the switch and bring the head of the nail close to your bar magnet. What do you observe?

____

____

11. As you discovered, in order for an electromagnet to have a magnetic effect (or field), a current must be passing though the wire. You also determined that the number of turns, or the length of coiled wire, affects the strength of your electromagnet. What other variables or factors might make your electromagnet stronger?

____

____

12. Design an experiment to test one or more of the variables that might make your electromagnet stronger. Be sure to get approval from your teacher before doing your experiment.

____

____

____

____

____

Will the True Colors Please Rise?

Colorful Experiments in Chromatography

Jacinta Martinez, S.N.D.
Sacred Heart School
Saratoga, California

I began teaching 29 years ago in San Jose, California, and have since taught in a number of schools throughout California and Hawaii. In 1968 I was assigned to teach science, and I "inherited" a science program from the previous teacher. At the time I didn't know anything about science, but I learned by doing and have now started my 20th year of science teaching in the middle grades.

Science fairs are a special interest of mine, and for the last 20 years my students have participated in local science fairs. In 1986 I was the recipient of the Outstanding Teacher Award from the Board of Directors of the Santa Clara Valley Science and Engineering Fair. In addition I have served as a member of Science Curriculum Committee for the Diocese of San Jose for the past three years, and as a member and past secretary of the Elementary School Science Association of Northern California for the past eight years. As part of my interest in environmental science, I directed the Montezuma Ecology Center in Los Gatos, California, for three years. Nearly 5000 students participated in the environmental science program sponsored by the Ecology Center, and I also conducted in-service workshops for teachers through the Center.

Helping other teachers who are starting out is important to me, just as I was helped when I first started. I was fortunate to learn the teaching craft from master teachers, and through my love of teaching I try to generate a love of science and an enthusiasm for learning in my students. Together they discover the joy of learning, as well as the excitement of exploring the next wonder in science and technology.

Art for this unit was provided by one of my students, Jane Dryden.

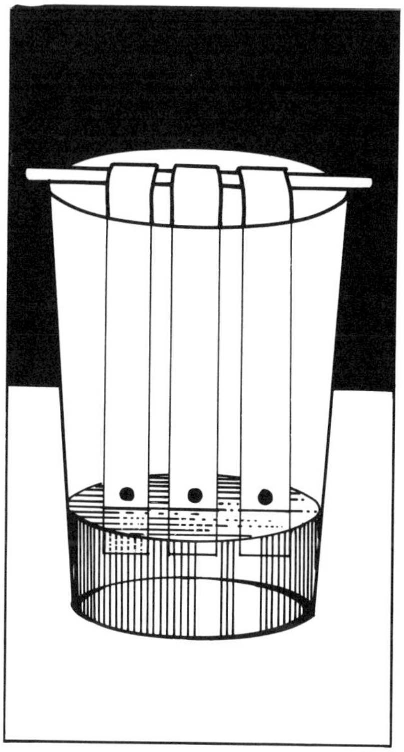

Lessons:

1. "Mix the Scientific Method and Chromatography"
2. "Kool-Aid Chromatography"
3. "Mystery Pens"

Overview

COLOR IS ALL AROUND US. It delights our eyes in a variety of ways—as fiery autumnal displays in forests and tree-lined neighborhoods, as the awesome arches of a rainbow, and as the brilliant pink cotton candy at the fair. Color can also be used to tell us about the substances that are present in a mixture through the process of chromatography.

The term *chromatography* is derived from the Greek words *chroma,* meaning "color," and *graphia,* or "writing." You might say that doing experiments with chromatography is like writing with colors, since the resulting chromatograms are like color signatures of different substances. The colors we see are but a tiny part of the spectrum of light that comes from the sun. Sunlight can be divided into two parts: visible light and invisible electromagnetic waves. Both parts make up the *electromagnetic spectrum.* Light travels in a very rapid wave-like motion. On striking the eye, signals are transferred in an instant to the brain, where we perceive these waves as color. It is important to know three simple facts about light: it travels in waves, these waves are of different lengths, and the wavelength determines the color we see.

One of the best examples of the visible spectrum is the rainbow. Light waves of different lengths travel through water droplets in the atmosphere, which act like miniature prisms. (A prism bends sunlight so we can discern the colors of the different wavelengths.) These waves are then separated, resulting in the rainbow. The shorter waves produce the violet shades; as the waves get longer we see blue, green, yellow, and orange, with the longest wave appearing as bright red. As this phenomenon shows, sunlight is a combination of all colors. You can demonstrate this to your students by making a color wheel and then spinning it to see if they can perceive white—the combination of all colors.

A second example of the visible spectrum can be demonstrated by way of the question "What gives objects their color?" Most objects appear colored mainly because of their chemical structure. A bright red balloon, for example, has a particular chemical make-up. When the balloon is bombarded with numerous light waves, these waves are either absorbed or reflected by the balloon. The reflected waves bounce back to us and are converted by the eye and brain to a vibrant perceived color. The color white is made when all the wavelengths of light are reflected, while the color black is made when all the wavelengths of light are absorbed by an object.

Visible light is only a minute section of the electromagnetic spectrum. The electromagnetic spectrum also includes gamma rays, x-rays, ultraviolet rays, visible light, infrared rays, radar, television, and radio waves (see Figure 1 below).

Although most of these wavelengths are invisible to us, our lives are affected by them every day. Ultraviolet rays tan us, and x-rays are useful in medicine. The remote control for your television uses infrared rays, airplanes are guided by radar systems, and the media world relies on television and radio waves. Electromagnetic waves get around! Studies in the field of optics include understanding how light interacts with objects such as mirrors, glass, and prisms. Chemists use the properties of light to analyze various compounds. Astronomers use their knowledge to get more information about objects in space.

The lessons found in this unit will show your students how to use chromatography and the properties of light to separate and identify substances in a mixture. This can be easily done in the classroom by placing a mixture onto an absorbent material such as filter paper and then using a solvent such as water or vinegar to begin the separation. The colors that rise up the filter paper indicate what substances are in the mixture. This color "record" is called a *chromatogram.*

This information sets the stage for you and your students to begin some fun experiments in chromatography. Your students will enjoy being chemists as they make the true colors rise up a chromatogram.

Key Concepts

- ☐ The scientific method is used to conduct investigations and experiments in a methodical manner.
- ☐ Light waves are either absorbed or reflected, depending on their wavelenth, thus determining the colors we see.
- ☐ The compounds in a mixture can be separated and identified through the process of chromatography.
- ☐ Dyes are colored substances that stain, but can be dissolved in liquids or separated from pigments.

gamma rays	x-rays	ultraviolet rays	visible light	infrared rays	radiowaves radar television radio

shortest waves ———————————————— longest waves

Figure 1. The electromagnetic spectrum.

Skills Used in the Lessons

- observing
- comparing
- predicting
- organizing
- measuring
- group cooperation
- communicating
- making conclusions
- using the scientific method

Extensions and Sources

Students interested in pursuing the study of color can do further research on how colors affect moods, and how they are used in fashions, interior decorating, paints, and chemical analysis. Colors are abundant in nature, and students can also research a wide variety of topics in this field.

In listing the resources I used in creating these lessons, I must credit the inspiration for Lesson 2 to an article by Christie L. Jenkins titled "Kool-Aid Chromatography," which appeared in the April 1986 issue of *Science and Children*, vol. 23, no. 7, pages 25-27.

The following books were also helpful in preparing this unit.

Minnesota Environmental Sciences Foundation, Inc. *A Teacher's Preparation for Color and Change: An Environmental Investigation into Magic Color in Nature.* Washington, D.C.: National Wildlife Federation, 1972.

Sourcebook for the Biological Sciences. Second Edition. San Francisco: Harcourt Brace Jovanovich, Inc., 1966.

Glossary

Absorbent: A material that absorbs matter, such as a liquid or a gas.

Chromatogram: The filter paper on which a solute is developed.

Chromatography: The process of separating substances in a mixture.

Compound: A substance containing two or more elements that are chemically combined.

Dye: A colored substance that has a staining effect, but can be dissolved in liquids or separated from pigments.

Insoluble: A mixture that cannot be dissolved.

Mixture: Two or more substances that are blended but not chemically combined.

Pigment: A colored compound absorbed by other materials to produce that color.

Reflection: The bouncing of a wave or ray off a surface.

Scientific method: A methodical way of investigating a problem or question.

Shade: The degree of darkness in color.

Soluble: A mixture that can be dissolved.

Solution: Two or more substances that are mixed and evenly distributed throughout.

Solvent: The substance in which a solute is dissolved.

Spectrum: The band of colors produced when sunlight passes through a prism.

Tint: A color made lighter when diluted with white.

Tone: The color that results from adding gray to any color.

Lesson 1

Mix the Scientific Method and Chromatography

An introduction to chromatography

Group Size
2-3 students

Time Required
1-2 class periods

Materials
For optional scientific method experiment:
- 3 test tubes and stoppers
- bottle of vinegar
- box of baking soda
- metric measuring tape
- measuring spoons
- safety goggles

For chromatography experiment:
Each group should have
- 1 bottle each of red, blue, and yellow food colorings (I use Schilling's because of their drop control vials)
- 1 clear plastic 4 oz/100 ml cup
- 1 clear plastic 8 oz/250 ml cup
- several pieces of filter paper or large cone coffee filters
- 1 metric ruler
- scissors
- clear cellophane tape
- 1 plastic drinking straw
- water
- paper towels

Each student should have
- several toothpicks
- pencil
- safety goggles (optional)
- copy of Handout 1, "Chromatography Procedure"
- copy of Handout 2, "Our Chromatography Experiment"

Key Terms

tint	chromatography
shade	scientific method
tone	dye

Instructional Goals
- To perform an introductory experiment that demonstrates the steps of the scientific method.
- To allow students to experiment with mixing and combining food coloring dyes to create various colors.
- To separate these color mixtures into their individual dye components using chromatography and the scientific method.

Student Objectives
Students will:
- Predict the food coloring combinations that make up different "mystery mixes."
- Set up and perform an experiment using the steps of the scientific method.
- Prepare and develop several chromatograms.

Prerequisite Knowledge
Students should know that red, yellow, and blue are the primary colors, and that these colors can be mixed to make any other color. Students should also be able to use a graduated cylinder and eyedropper.

Advance Preparation Time
About 30 minutes
- Review the "Background Information" and "Lesson Procedure."
- Gather materials.
- Duplicate student handouts.

Teacher Tips

- Make sure your students understand this fundamental lab rule—chemicals or solutions used during experiments are *never* to be tasted except by permission of the teacher. The food colorings in this experiment are not toxic, but this rule is a good one to follow as a basic lab procedure.
- Although food coloring is not an eye irritant, I usually have my students wear safety goggles as a precautionary measure.
- If possible, set up two or three stations for distribution of materials. This makes the distribution process easier and faster.
- You may find that using six chromatograms for each group works better than ten, especially if your class tends to be slow when performing lab experiments.
- I find that if I'm personally enthusiastic about an experiment or lesson, my students also pick up this sense of enthusiasm, and the discovery process becomes a cooperative venture.

Background Information

What do the Grand Canyon, a peacock's tail, red apples, and a bouquet of flowers have in common? Colors!

To begin the study of colors and chromatography, your students will need to start with the *primary colors*—red, blue, and yellow. The primary colors can be combined to give us an infinitesimal array of colors. A rainbow shows us the separation of light in its many colors. White light is a combination of all the colors. A *tint* is produced when white is added to a color. Black added to any color will produce a *shade,* and gray added to any color makes a *tone. Pigments* are colored compounds in plants or paints that produce a color. *Dyes* are colored substances that have a staining effect, but can be dissolved in liquids or separated from pigments.

These concepts will be easier for your students to understand when they mix simple food coloring dyes in this activity and create color "magic" right before their eyes. They can then separate these dyes through paper chromatography.

Chromatography is the process of separating substances in a mixture (such as the different food colorings in this experiment). In paper chromatography, this is done by placing filter paper that has been stained with a drop of the mixture into a solvent (water) so the various substances can be separated and identified. Chromatography is colorful, and your students will enjoy watching the separation of these colored dyes as they migrate up the filter paper. At the same time, your students will use the step-by-step procedure of the scientific method.

The scientific method is the framework used by scientists to carry out an experiment in a methodical way. It helps your students state a problem and hypothesis, predict the outcome of an experiment, perform the experiment, reach conclusions, and make suggestions for further experiments to verify their conclusions. You'll find that when your students use the scientific method, their papers will show you not only their thoughts, methods, and procedures, but also will give you a better way to evaluate their conclusions.

Lesson Procedure

For Day 1 (an optional experiment that introduces the scientific method)
Step 1:
In this experiment, students will mix vinegar and baking soda inside a test tube, cork the test tube, and then measure how far the cork flies due to the carbon dioxide build-up inside the test tube. I've found this simple experiment is very useful for introducing the steps of the scientific method to my students.

The first step of the scientific method is to state a problem or question. For this experiment, a simple question could be, "How far will carbon dioxide in a test tube shoot a cork stopper?"

Step 2:
Formulating a hypothesis is the next step of the scientific method. This is an educated guess that predicts how the experiment will turn out. For example, a hypothesis for this experiment could be, "If carbon dioxide is an expanding gas, then the cork on the test tube will shoot out." This is called the "If...then" method, which is sometimes difficult for students to understand, but becomes easier with practice.

Step 3:
When beginning to learn the scientific method, it is a good idea for students to list materials they used. Once students have mastered this step it can be omitted or replaced with quick illustrations.

Step 4:
The next step of the scientific method is to list the experiment's procedures, which should resemble a step-by-step list of directions. Once mastered, this step can be omitted or stated briefly. For this particular experiment, one student should pour 1/8 teaspoon of baking soda into a test tube. The class should then go outside (bringing along the rest of the materials), and the student with the test tube should stand at a starting line designated by you. Have two students stand near the student with the test tube—one on each side. One student will pour 5 ml of vinegar into the test tube, then the other student should quickly cork the test tube with a stopper. The test tube should be aimed toward an open field (such as a football or softball field), and two additional students should stand by to measure the distance of the cork's flight. After the cork goes off, one student should stand at the position where the cork landed, while the other student measures the distance the cork traveled. The remainder of the class stands to the side, cheering them on and recording the results. Note: Student volunteers should wear goggles.

Step 5:
Recording observations and results is an important part of any experiment. Students should keep accurate records of the experiment's results, because these records are the primary data used to test the hypothesis. As part of their records students can also report any successes or problems that occur. Results can be noted in tables, illustrations, or graphs (see Figures 1 and 2 on page 181).

Step 6:
The conclusion is the final step of the scientific method, and it ties together the entire experiment. The original question and hypothesis will be answered in the conclusion, depending on the results of the experiment. If the experiment fails, students should note possible reasons why and what further experiments could be done.

For Day 2 (Chromatography)
Step 1:
I begin the chromatography experiment by asking students what their favorite colors are. Then I tell them that in today's lesson they will enjoy doing

three things: creating colors, writing about their experiment in the same way scientists do (using the scientific method), and separating the colors they create into their various components through the process of chromatography.

Step 2:
Divide your students into groups of two or three and assign a number to each group. One member from each group should pick up the group's materials. Have students experiment with mixing primary colors. (If you are using Schilling's food colorings, have students use the formulas given on the back of the box.) To practice for the next step, students can use their toothpicks to place dots of their color mixtures on a paper towel. After a few minutes of this, have students rinse out their cups and clean up. The cups will be used again in the next step.

Step 3:
I like to introduce this part of the experiment by demonstrating how to set up a chromatogram. Pour water into a 250 ml cup to a depth of 1 cm (or 20 mls). Cut a strip of filter paper 12 cm by 1 cm. Use a toothpick to place a dot of food coloring on the filter paper 1.5 cm from the bottom (see Figure 3 on page 182). Roll the filter paper over a drinking straw and suspend the chromatogram in the cup so that the dot of dye is just above the water level (see Figure 4 on page 182). Use tape to fasten the chromatogram to the straw. In this experiment, each group will make a "mystery mix" of colors, write down their formula, and then make several chromatograms using their mystery mix. Students will then swap chromatograms, develop them, and try to figure out the different food coloring combinations.

Trial 1	Trial 2	Trial 3
The cork was not put in fast enough and the CO_2 fizzled out – 0 meters.	Success! Cork went 7 meters.	Cork flew 8 meters.

Figure 1. A sample table of results from the shooting cork experiment.

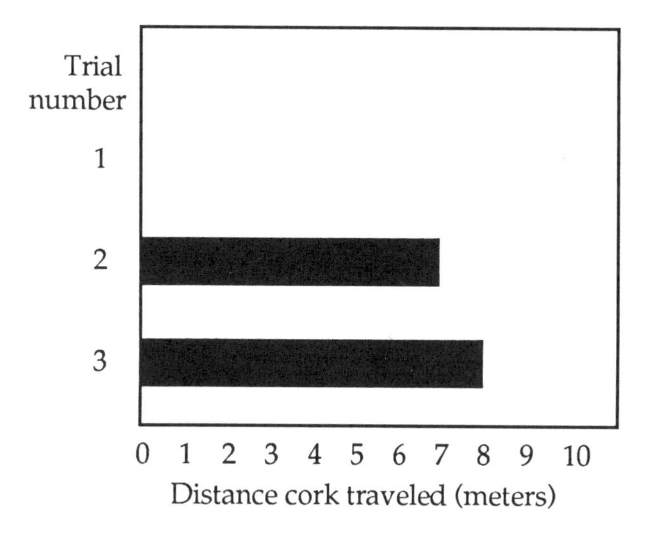

Figure 2. A sample graph of results from the shooting cork experiment.

Step 4:
Distribute the student handouts, and have students read Handout 1, "Chromatography Procedure." Then on Handout 2, "Our Chromatography Experiment," students should write a title, hypothesis, and list of materials. As they proceed with the experiment, students should write their observations and conclusions on Handout 2. Circulate from group to group during the experiment to answer questions. If time permits, students can also compare their findings in a class discussion. As part of the discussion, I like to ask students these questions:

- Did the amount of solution used for the color dots on the various chromatograms seem to make any difference in your results?
- Did some other color show up on the chromatogram that was not in the mixture? Why or why not?
- Was it helpful to follow the steps of the scientific method?
- Can you do a scientific experiment without proposing a question or hypothesis? Without explaining your conclusion?
- Can you think of other uses for chromatography?

Enrichment Activities

- ☐ Make a "color corner" in your classroom where students can use food colorings to make other color combinations. These combinations can be recorded on a class color chart.
- ☐ Make a color bulletin board. Examples for themes include colorful birds, cars, clothes, etc., or paint chips of various tints and shades.
- ☐ Students can do research on plants and vegetables that are used to make color dye solutions, and then bring samples to class.
- ☐ Questions for students to research:
  - How is color used by animals for protection or to attract mates?
  - Are all leaves on trees and plants truly green? How many shades or tints of green can you find?
  - What is meant when people "have their colors done?" What colors are best for you to wear?

Figure 3. A dot of food coloring is placed 1.5 cm from the bottom of the filter paper strip.

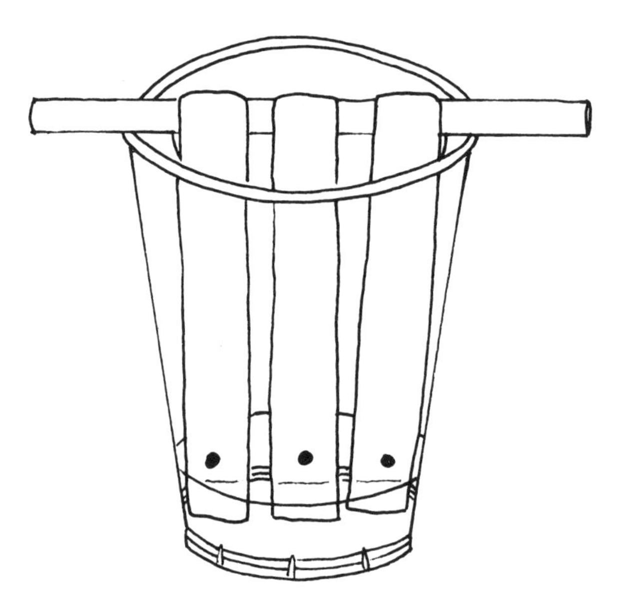

Figure 4. Three chromatograms are fastened to a straw and placed in a cup so that the food coloring dots are just above the water level.

Chromatography Procedure

1. Using your small plastic cup, prepare a "mystery mix" of food colorings. Use only the primary colors—red, blue, and yellow. You can mix any or all of these colors in any combination you like. Write the formula you used on a piece of paper; for example, 2R + 1B + 15Y. Label this paper with your group number and give it to your teacher. This formula will be used by another group to confirm their conclusion.

2. Cut ten strips of filter paper so that each one measures 12 cm by 1 cm. Using a toothpick, place a round dot of your mystery mix solution on each strip of filter paper 1.5 cm from the bottom. With a pencil, write your group number on each strip below the dot.

3. Take your ten strips to the place designated by your teacher. Now you are ready to exchange strips with other groups. Choose ten strips, or *chromatograms,* from different groups. You will use the process of chromatography to figure out the colors that make up each mystery mix.

4. Fill your 250 ml plastic cup with water to a depth of 1 cm (or 20 mls). Attach each chromatogram to a straw as shown in the drawing below, and place it in the cup. It is very important for you to adjust each chromatogram so that the colored dot is *above* the water level.

5. Wait for the chromatograms to develop. Write your observations on Handout 2. When the water has reached the top of the chromatograms, they are developed. Remove the chromatograms from the cup and allow them to dry. Clean up while you're waiting, and write your conclusions on Handout 2 about what colors were used to make up each mystery mix.

6. When you've completed your conclusions for each chromatogram, compare your results with the original formula. On Handout 2, describe how close your identification came to the actual formula. Be sure to answer these questions:
   - Was your original hypothesis correct or incorrect?
   - Did your chromatogram help you test your hypothesis?
   - What further experimentation would you like to do?

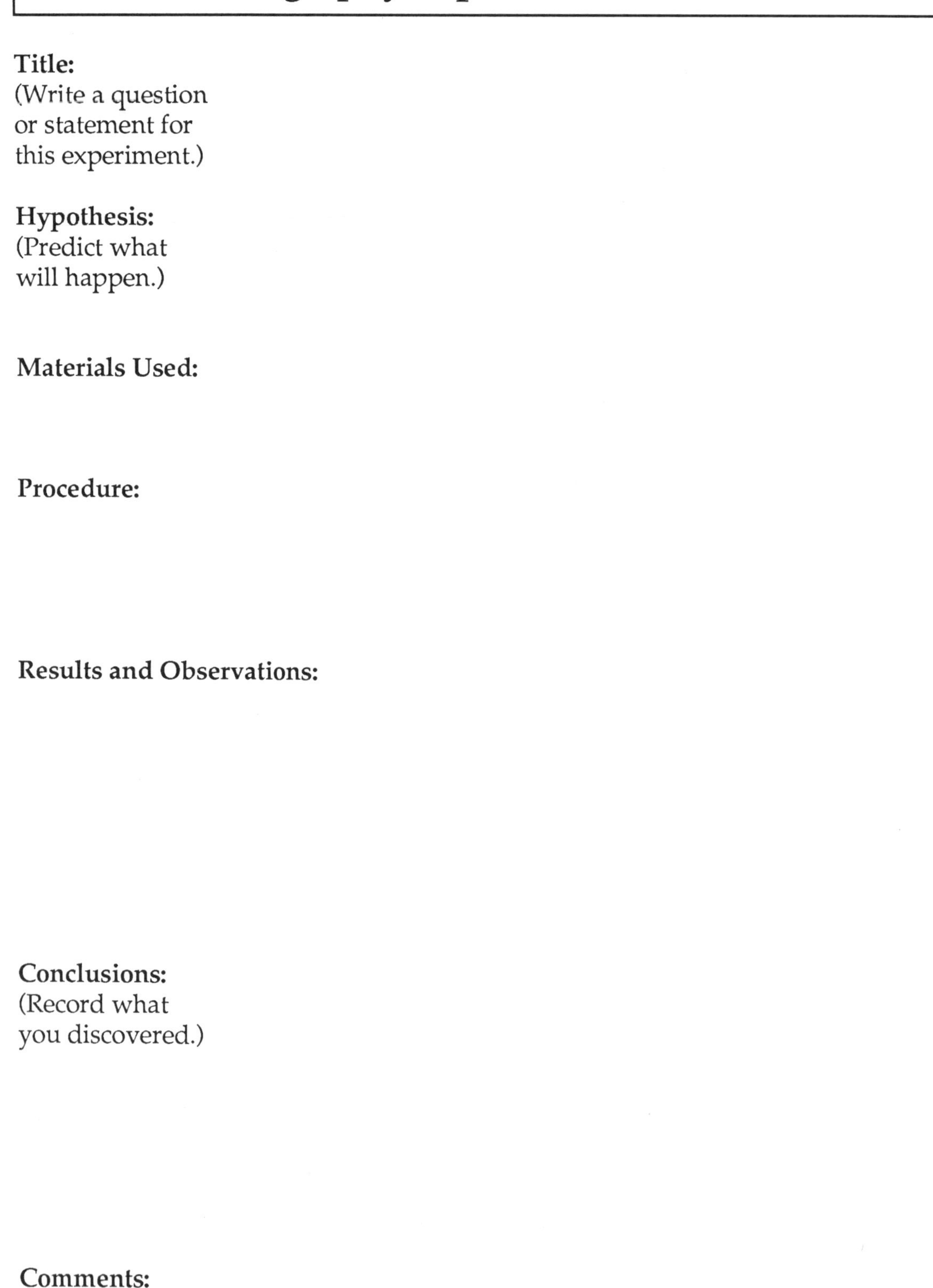

Our Chromatography Experiment

Title:
(Write a question
or statement for
this experiment.)

Hypothesis:
(Predict what
will happen.)

Materials Used:

Procedure:

Results and Observations:

Conclusions:
(Record what
you discovered.)

Comments:

Lesson 2

Kool-Aid Chromatography

Separating dyes used in food

Group Size
2 students

Time Required
1 class period

Materials
- 6 packets of assorted flavors of unsweetened Kool-Aid
- 6 clear plastic 8 oz/250 ml cups
- 1 small package of M & M candy (optional)

For each group:
- 1 clear plastic 8 oz/250 ml cup
- 1 plastic drinking straw
- several sheets of filter paper or large cone coffee filters
- water
- graph paper
- clear cellophane tape
- scissors
- pens and pencils
- colored markers or pens

For each student:
- several toothpicks
- safety goggles (optional)
- copy of Handout 1, "Kool-Aid Chromatography Procedure"
- copy of Handout 2, "Our Kool-Aid Experiment"
- copy of Handout 3, "Kool-Aid Dyes"

Key Terms

chromatography	mixture
spectrum	compound
soluble	absorb
solution	reflect
solvent	dye
tint	tone

Instructional Goal
- To discover that mixtures in everyday substances, such as food dyes in beverages, can be separated and identified through the process of chromatography.

Students Objectives
Students will:
- Separate the dyes found in Kool-Aid.
- Identify the dyes and note the results using the scientific method.

Prerequisite Knowledge
Students should be familiar with the material in Lesson 1, particularly the procedure used for paper chromatography and the steps of the scientific method.

Advance Preparation Time
About 30 minutes
- Review the "Background Information" and "Lesson Procedure."
- Gather materials.
- Duplicate the student handouts.

Teacher Tips
- Students should know that, as a rule, they are *never* to taste solutions used in an experiment unless they have your permission. In this lesson, the Kool-Aid solutions are harmless. For fun, have student volunteers taste the unsweetened concentrate solutions and note their reactions.
- Set up two or three stations of materials to facilitate distribution.
- Safety goggles are always recommended for experiments, even though students are not working with toxic substances in this activity.

Background Information

Are you or your students aware of the dyes that are added to the munchies and drinks you consume every day? Food colorings are added to foods for one basic reason—to make them more appealing to consumers. Many food labels state, "Artificial coloring added." Because a portion of the population is allergic to yellow dye number 5, this dye is required to be listed on food labels when it has been added. But what are the other food colorings that are commonly added?

Before beginning this lesson, review with your students the concept of visible light and how the reflected wavelengths determine the colors we see. Also review the fact that pigments are colored compounds found in plants and paints, while dyes are substances that can be dissolved or separated from a pigment. In this lesson your students will investigate some of the colored dyes found in a common beverage (Kool-Aid) through the simple process of chromatography, which will separate the dyes so they can be identified. Students usually enjoy watching the migration and separation of the colored dyes. As they do this experiment, students should be aware that the rate at which a mixture rises up a chromatogram depends on the chemical make-up of the mixture, as well as the properties of the cellulose fibers of the filter paper.

As part of this lesson I like to explain that chemists use several different types of chromatography—liquid, thin layer, paper, and gas—to analyze many chemical compounds. Examples of compounds that are commonly analyzed include pigments, sugars, drugs, hormone levels, blood, urine, cells and tissues, and even pollutants in air and water.

Lesson Procedure

Step 1:
Empty six packets of assorted flavors of unsweetened Kool-Aid into six different clear plastic cups, and label each cup with the corresponding flavor. Add 30 ml of water to each cup and mix. This will produce concentrated solutions.

Distribute the student handouts, divide your students into groups, and have students pick up their materials.

Step 2:
In this experiment, students will use chromatography to find out what dyes are used in Kool-Aid. Begin the lesson by asking students to name a few additives found in the foods we eat; for example, sugar, artificial sweeteners, salt, artificial flavors, etc. To provide an example of why food colorings are used, I like to open a small package of M & M candy and ask students what makes the candy so appealing. Ask if any of your students are allergic to certain foods or food colorings. Tell them that some people are allergic to a food coloring called yellow dye number 5, and therefore this dye is required to be listed on food labels. Then show your students the concentrated Kool-Aid solutions you made.

Make sure students know that the dissolved Kool-Aid is the *solution* they will use for this experiment, and that water will be the *solvent* they use to develop their chromatograms. Explain that some substances are *soluble* and some are not. A substance is soluble if it can be dissolved in a solvent. Substances that cling or remain in the same position on a chromatogram are insoluble.

Step 3:
Have students begin their experiments, following the instructions on Handout 1, "Kool-Aid Chromatography Procedure." Note that students will cut

the filter paper into a rectangle, which is then rolled into a cylinder, attached to a straw, and suspended in a cup. Alternatively, you can have your students use individual strips of filter paper as they did in Lesson 1. If you use the filter paper cylinder, make sure students place their dots of solution 1.5 cm apart.

After the chromatograms have developed, students can compare their results with the chart on Handout 3, "Kool-Aid Dyes," which lists the dyes found in various flavors of Kool-Aid. For each flavor, the dyes will separate in the order they are listed. Students should write their observations, results, and conclusions on Handout 2, "Our Kool-Aid Experiment."

Step 4:
Have students compare their findings through class discussion. I use these questions as a basis for discussion:

- Which flavor(s) contained yellow dye number 5? Did the manufacturer list it on the packet?
- Which dye migrated the fastest? Which dyes were insoluble?
- Did different kinds of filter paper make a difference? Could you have obtained the same results by using a paper towel, tissue, or napkin?
- How would you explain chromatography to a younger student or your parents?
- Do you make a practice of reading food labels on the foods you eat? Do you think the general public reads food labels very often? Do you think that food labeling is necessary?

Enrichment Activities

- ☐ Make a "color corner" where students can perform other chromatography experiments based on their own hypotheses.
- ☐ Have students calculate the "Rf" value for each chromatogram, which stands for the "rate of flow." Use this formula:

$$Rf = \frac{\text{distance solute traveled}}{\text{distance solvent traveled}}$$

- ☐ Students can do research on fabrics that are color-fast. What makes some materials color-fast while others are not?
- ☐ Students can explore other dyes found in foods, drinks, and candy. As part of this research, students can write to manufacturers for information on dyes and additives found in selected products.

Kool-Aid Chromatography Procedure

1. Pour water into a plastic cup to a depth of 1 cm (about 20 mls). Then cut a rectangle of filter paper about 12 cm by 10 cm. With a pencil, draw a straight line across the width of the rectangle 1 cm from the bottom of the paper. Using the tip of a toothpick, place a small dot of each Kool-Aid solution on the line, leaving a space of at least 1.5 cm between each dot.

2. Form a cylinder with the filter paper, and pass a plastic straw through two slits cut into the sides of the cylinder (see Figure 1). You will need to adjust the size of your cylinder to the size of your plastic cup. (Alternatively, you can cut the filter paper rectangle into individual strips.) Suspend the filter paper so it does not touch the sides of the cup, and so that the bottom of the paper touches the top of the water level. The dots of dye *must* remain above the water level (see Figure 1), otherwise your water will be contaminated and you will have to start over.

3. Wait for your chromatogram to develop. When the water has reached the top of the paper, your chromatogram is finished developing. Remove the paper and allow it to dry. Clean up while you're waiting, and write your observations and results on Handout 2. Then check your results against the "Kool-Aid Dye Table" on Handout 3. Note how your results compared with the table.

4. Construct a chart, table, or graph showing a comparison between the dyes that were soluble and those that were insoluble. Use graph paper for charts or graphs. You may also want to include in your results any shades, tints, or other colors that surfaced in your experiment. Include your chart, table, or graph in the "Results" section of Handout 2. Then write your conclusions.

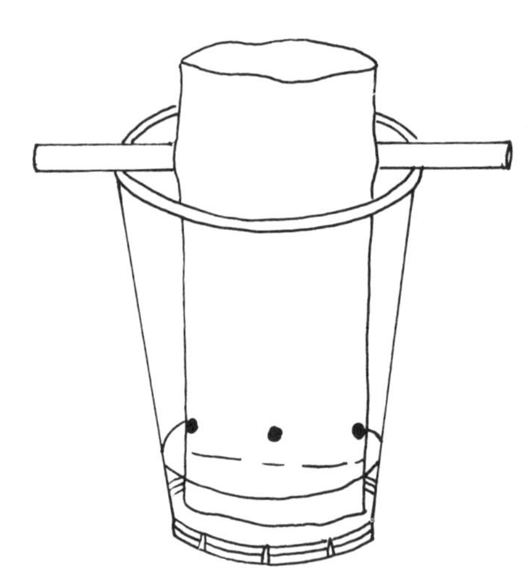

Figure 1. Roll the rectangular filter paper into a cylinder, pass a straw through it, and place it in the cup so that the dots are above the water level.

Our Kool-Aid Experiment

Title:
(Use a "colorful" title or ask a question.)

Hypothesis:
(Predict what will happen.)

Materials:

Procedure:

Results and Observations:

Conclusions:
(Record what you discovered.)

Kool-Aid Dyes

Listed below are the dyes that make up each flavor of unsweetened Kool-Aid. The dyes will separate in the order listed, from top to bottom, for each flavor.

Unsweetened Kool-Aid Dye Table*

Flavor	**Dyes**
Apple	Blue Number 1 Red Number 40 Yellow Number 6
Black Cherry	Blue Number 1 Red Number 40
Cherry	Blue Number 1 Red Number 40
Grape	Blue Number 1 Red Number 40
Lemonade	Yellow Number 5
Lemon/Lime	Blue Number 1 Yellow Number 5
Orange	Red Number 40 Yellow Number 5
Pink Lemonade	Red Number 40
Rainbow Punch	Red Number 40 Yellow Number 5 Yellow Number 6
Raspberry	Blue Number 1 Red Number 40
Strawberry	Red Number 40
Strawberry Falls	Red Number 40 Yellow Number 6
Sunshine Punch	Yellow Number 6

---

* From Christie L. Jenkins, "Kool-Aid Chromatography," *Science and Children* 23, no. 7 (April 1986): 27.

Lesson 3

Mystery Pens

The Case of the Anonymous Admirer

Group Size
2 students

Time Required
1 class period

Materials
- 6 or more ballpoint or felt tip pens of assorted brands; black and blue usually work well
- a note written on a large sheet of filter paper with one of the pens, cut into as many chromatogram strips as you have student groups
- one business-size envelope
- vinegar

For each group:
- 6 strips (12 cm by 1.5 cm) of filter paper or cut from large cone coffee filters
- 1 clear plastic 8 oz/250 ml cup
- 1 or 2 plastic drinking straws
- several toothpicks
- tape and pencils

For each student:
- copy of Handout 1, "Mystery Pens Procedure"
- copy of Handout 2, "Our Mystery Pen Experiment"

Key Terms

solvent	chromatogram
soluble	dye
insoluble	pigment

Instructional Goal
- To use the process of chromatography to analyze a variety of ink samples, leading to the identification of the writer of an anonymous note.

Student Objectives
Students will:
- Analyze and compare inks from a variety of pens.
- Use the scientific method and chromatography to separate and identify ink samples.

Prerequisite Knowledge
Students should understand the concepts and procedures of chromatography and the scientific method as introduced in the first two lessons of this unit.

Advance Preparation Time
About 30 minutes
- Review the "Background Information" and "Lesson Procedure."
- Gather materials.
- Make copies of the student handouts.

Teacher Tips
- I have found that vinegar works faster than water as a solvent for this experiment.
- Be sure to use clear plastic cups rather than opaque ones so that students can more easily see their chromatograms develop.
- Safety goggles are always recommended for laboratory experiments, even though students are not working with toxic substances in this activity.
- You may use pens of different brands that have approximately the same color ink, since their chromatograms may differ.

Background Information

One of the things that amazes me is the spectrum of colors that I see in my local stationery store, with its variety of pens, felt markers, and pencils. The combination of colors is delightful, and I can't walk down the aisle without trying out at least some of them.

The ink in most of these pens and markers is made up of pigments that can be separated through chromatography. The color that we perceive is dependent on the chemical make-up of the ink, which reflects light of a certain wavelength. When our eyes see this wavelength, our brains convert it into a vibrant perceived color.

If your students have enjoyed the two previous lessons on color chromatography, they will enjoy this lesson as well. Through this experiment they will discover that the colors of ink in ballpoint pens are not what they appear to be. Does a black ballpoint pen write only in black ink? Is a green pen really green? Use chromatography to find out.

Lesson Procedure

Step 1:
Label each of the six ballpoint pens with fictitious names, or use names of famous people. Using a large piece of filter paper and one of the pens, write an "anonymous" note. Cut the note into strips about 12 cm by 1.5 cm. You'll need at least as many strips as you have student groups. Then place all the strips into an envelope.

Step 2:
Motivate your students by telling them that they have received an anonymous note from a "secret admirer." Fortunately, they can use chromatography to find out who wrote it. Their assignment is to act like detectives by analyzing and comparing the inks in each of the six sample pens to find the author of the note.

Distribute the student handouts and materials, then have students follow the instructions on Handout 1, "Mystery Pens Procedure."

Step 3:
When students are finished developing chromatograms for each pen, have each group take one strip of the "mystery note" from the envelope and repeat the chromatography procedure with this strip. Remind students that as they discover which pen wrote the note, they should keep their answer a secret until the whole class is done. My eighth graders loved this experiment, and could hardly wait to compare results.

Step 4:
After all the groups have finished, open the class discussion by asking each group to name the author of the mystery note. Include these questions in the discussion:

- Did you find any pigments that were insoluble?
- Which colors rose at a steady rate? Which were slow? Why?
- Which pens had more than two color components? What primary or secondary colors were in these inks?
- If the secret admirer had written the note with two different pens, would you still have been able to discover which pens were used? Explain your answer.
- What problems did you experience? How could you have made the experiment better?

Mystery Pens Procedure

1. Use each of the six pens to write a large round dot on a strip of filter paper. Place each dot 1.5 cm from the bottom of the paper.

2. Pour 20 ml of vinegar into your plastic cup. Wrap each filter paper strip around a plastic drinking straw. Adjust the strip so that when you suspend it in the cup it will just touch the vinegar, then tape the strip in place. The dot of ink must be *above* the vinegar level (see Figure 1).

3. Watch the colors separate and migrate up the chromatogram. This may take a while. As colors leave the dot, they may look like miniature comets on your chromatogram. Record your observations on Handout 2. When you think the colors are no longer migrating, set the chromatogram out to dry.

4. When all your chromatograms have dried, write your results on Handout 2. Now you are ready to get a sample of the mystery note from your teacher. Use the same chromatography procedure to develop it. With this sample, you will be able to compare colors and identify the pen that wrote the note.

5. Write the name of the author of the note in the "Conclusion" section of Handout 2. Keep your conclusion secret until your teacher asks for your answer during the class discussion, when you will see if your answer is correct.

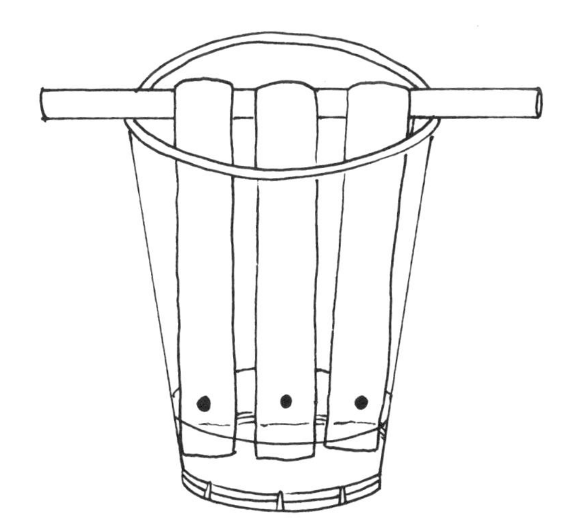

Figure 1. Place the strips in your cup so that the dots are above the vinegar level.

Our Mystery Pen Experiment

Title:
(Write a question
or statement for
this experiment.)

Hypothesis:
(Predict what
will happen.)

Materials:

Procedure:

Results and Observations:

Conclusion:

Comments:

Appendix
Computers in the Science Classroom

Judith Scotchmoor
Marin Country Day School
Corte Madera, California

In 1968, with sweaty palms and shaking knees, I faced the challenge of first-year teaching. I survived and discovered how to communicate effectively with students not much younger than myself. Following that experience and additional study in Sweden, I taught for five years in a boys' prep school in England. The role of a teacher in England is that of an ally—an essential key to a child's personal academic goals and a partner in the learning process. My confidence grew and I began to understand that each child is unique and places different demands on me—as a tutor, mentor, disciplinarian, "parent," and friend. Also during this time in England I learned that the higher I set the academic goals in my classes, the more my students achieved.

I returned to California with a broader outlook on education and began teaching seventh and eighth grade math and science. My professional growth continued as I became interested in using computers in education. This interest was sparked in the late 1970s when I decided I should invest in a calculator and casually went shopping at a local consumer electronics store. While waiting for assistance, I started watching someone typing at one of the computers on display. Looking up, I spotted a sign that advertised the store's free computer training course for teachers. The price was right, so I began my personal computer education.

My curiosity lead me to visit other schools that were already using computers to see how these tools were being used in their classrooms. What I saw was exciting and full of potential, so I asked my principal for permission to start a fund-raising campaign to buy a computer for our school. The parents and students were so supportive that we surpassed our goal three times. By 1986 the school had seven computers and two printers in a computer lab, as well as a computer in every classroom.

In addition to teaching at Marin Country Day School, I have also taught an extension course for teachers at the University of California at Berkeley for the last four years entitled "Computers in the Learning Process." My experiences have convinced me that computers are a very effective tool in classroom problem-solving situations and in teaching critical thinking skills.

Janet Graeber
Convent of the Sacred Heart School
San Francisco, California

After earning my bachelor's and master's degrees in education from the University of Virginia, I taught in elementary schools in Virginia, Massachusetts, Texas, Maryland, and the Bahamas before coming to San Francisco in 1982. Currently I am the computer coordinator at Convent of the Sacred Heart Elementary School and Stuart Hall for Boys. Student instruction is only part of this multi-faceted position; I am also responsible for curriculum development, teacher training, grant writing, and fiscal management. I have also taught a course at the University of San Francisco entitled "Instructional Uses of Computers."

My interest in using computers in the classroom has lead me to conduct several workshops and to act as a computer consultant to a number of schools in California. Computers are wonderful tools for teaching problem solving, writing, science, social studies, and history. As the number of software applications and hardware tools increases each year, educators will find more and more possibilities for integrating computers into their classrooms.

IT SEEMS INEVITABLE that computers would become a vital part of the science classroom. A computer is a tool, and tools have always been used by scientists to perform experiments and verify hypotheses. The increased sophistication of high-technology tools has enabled scientific discoveries to advance at an ever-faster pace. Of course, many sophisticated scientific tools are beyond the reach of K-12 teachers due to high costs or difficulty in using them. Not so with computers. Computer prices have dropped dramatically in the last several years, and the proliferation of different personal computers allows you to choose the one that's right for your budget and needs. The number of software applications has also increased substantially, and many easy-to-use programs have been written and marketed for use by educators and their students.

A computer can be part of an exciting learning center in your classroom. It can be an effective tool to:

- introduce, develop, and reinforce critical thinking skills
- bridge the skills learned from one area of the curriculum to other areas
- simulate laboratory experiments safely
- gather quantitative information using plug-in scientific tools, such as temperature and light sensors
- store and organize quantitative information
- provide evaluation activities

In this appendix we have listed a variety of software programs that have been used successfully in our own classrooms. This is by no means a comprehensive list of available educational software, but it is provided as a starting point for integrating computers into your science curriculum. Prices quoted are list prices at the time of printing.

High Wire Logic
Sunburst Communications
39 Washington Avenue
Pleasantville, New York 10570
(800) 431-1934 or (914) 769-5030

Available for Apple II family, IBM PC and IBM PCjr
Color monitor required
$65.00

A study of classification systems involves the recognition of critical and variable attributes. Students learn to observe, compare, contrast, and organize by category. The use of these skills is not confined to the science lab, however. While teaching your students to discriminate between a reptile and a mammal, for example, you are also simultaneously introducing or reinforcing critical thinking skills that are applicable throughout the curriculum. Consider the following classification comparison in science and language arts.

It has scales.
It is cold-blooded.
It lays eggs.
It must be a reptile.

It begins with a capital letter.
It contains a verb.
It ends with a punctuation mark.
It is a complete sentence.

A sentence has critical attributes, such as words and punctuation, just as a reptile does. Variable attributes, such as the scales of the reptile, help us classify it further. Similarly, the presence of key words or punctuation marks helps us identify a sentence type further. The ability to gather information through observation, organize it, analyze it, and then formulate theories is relevant and vital to all aspects of the curriculum.

Currently, there are a number of software products available that either introduce or further develop these skills. This particular program, *High Wire Logic*, takes students a step further by using Boolean logic statements (statements that show relationships using the words, "and," "or," and "not"). Through this program, students will be able to apply Boolean logic to the formulation of rules that distinguish one set of objects from another.

Excellent documentation is provided with the software. Included are ideas for introducing Boolean logic to your students, with accompanying transparency masters. As with all software, it is important to familiarize yourself with this information and the program before using it with your students.

When using the program, students are presented with two sets of figures that are generated randomly. One group of figures is positioned on a high wire, while the other group is positioned in a net under the high wire. The figures on the high wire belong together because of particular attributes, such as color, size, shape, or pattern. The figures in the net do not belong with the group on the high wire. The goal is to formulate as many rules as possible that will apply to the figures on the high wire and not to the other figures.

You will find that students become very involved when using this program. Several times in my classroom, students have been so intent in finding the rules for a particular grouping that we kept the computer on all day.

—J. S.

Discover
Sunburst Communications
39 Washington Avenue
Pleasantville, New York 10570
(800) 431-1934 or (914) 769-5030

Available for Apple II family
Color monitor required
$65.00

Nothing beats hands-on experimentation by students, but there are times when such activities are impractical due to expense, unavailable equipment, or safety considerations. A computer simulation can provide students with a simplified model of a laboratory experience while avoiding these drawbacks. Rather sophisticated experimental techniques can be safely tested, even with limited experience, and students can test a variety of conditions without damage to living organisms.

Simulations can also take a step into the future. This software program, *Discover,* is a computer activity that allows students to assume the role of a laboratory scientist. The simulated lab setting asks students to discover how to keep eight alien "organisms" alive and healthy.

As the simulation begins, Creature 1 enters the on-screen "laboratory" from its cryogenic pod. It moves around, bumping into walls and changing direction, apparently unconcerned about its new environment. Suddenly its color changes. Messages are displayed at the bottom of the screen that the creature is "suffering from acute anxiety attacks." Does it require food? If so, which food does it prefer? How does it eat? Does it need more room to maneuver? Is it lonely? Students must observe behavior patterns, determine food preferences, and recognize signs of anxiety for each specimen. As each creature is successfully maintained, new creatures can be added to the laboratory. Further observations must be made on their interaction with one another, as well as their individual needs, in order for these life forms to survive.

A couple of teacher tips:

- Become familiar with the software yourself by using it just as your students will. It is very difficult to support all eight creatures at the highest level of the simulation, but you will certainly experience the same excitement and frustrations as your students.
- After selecting "Begin a New Experiment" from the main menu, keep the disk in the drive until you see the message, "Now preparing the laboratory." The disk may then be removed from the computer and replaced with a data disk on which you can store lab results. Should you wish to return to the main menu, you must reinsert the master disk.

This is a great activity to encourage note taking and to improve observational skills, analysis, and the formulation of hypotheses. Best of all, this computer simulation allows you to demonstrate, without sacrificing any lab animals, that all creatures have certain needs and interact with each other in an environment.

—J. S.

Science Toolkit Master Module
Brøderbund Software
17 Paul Drive
San Rafael, California 94903
(415) 492-3500
Also available: Speed and Motion Module, Earthquake Lab, and Body Lab

Available for Apple II family
Joystick port or adapter required
$69.95

One of the primary objectives of science educators is teaching students the importance of making accurate observations. We encourage, if not demand, our students to take quantitative measurements at every possible opportunity; however, this is not always practical. In some experiments, particularly those dealing with light and heat, our labs are simply not equipped with tools sensitive enough to detect small changes in energy levels, or with tools capable of taking continuous measurements over long periods of time—enter the computer and the science tool kit.

There are several products of this type on the market, each with varying features and prices. My experience is with the *Science Toolkit* by Broderbund, which contains an easy-to-use temperature probe and light probe, complete documentation, explanations of scientific terms, and ideas for experiments. This kit allows your students to gather and organize quantitative data, formulate and test hypotheses, and understand important scientific concepts.

With this particular tool kit, a small interface box plugs into the back of the computer. A temperature probe (thermistor) and/or light probe (photocell) can then be connected to this box. Turn on the computer, place the disk in the disk drive, and your science lab is all set, complete with easy-to-understand, on-screen directions. While using the probes, your students will be able to read the data directly from the screen, and data will also be shown in a graph format. The results may be saved on another disk to be reviewed later.

A couple of teacher tips:

- Each brand of tool kit works a bit differently. Familiarize yourself with the one you purchase and read the documentation thoroughly before using it with your students.
- Encourage your students to repeat their experiments and average their results. This will compensate for possible human or mechanical errors in the collection of data.

Use the ideas for experiments provided in the product documentation, or add heat and light sensor experiments to the activities provided in this book. For example, after doing the lesson "Build Your Own Battery" on pages 147–154, have your students measure the light intensity (in foot candles) of the flashlight bulb that is connected to the student battery. This can be done by placing a cardboard tube over the light bulb and inserting the light probe, taking measurements over a 30-second time period to show how the light bulb dims as the battery runs down. Students can also compare the results of placing light bulbs in series and parallel circuits, or they can test a variety of situations by adding batteries, switches, and other devices.

After investigating color in the chromatography unit (pages 174–194), your students can use the light probe to measure light absorbance and reflection of different colored sheets of construction paper. By measuring the light intensity of the various colored papers, students can prove to themselves that light colors reflect more light, and dark colors absorb more light.

When setting up an aquarium as part of the ecosystems unit (pages 120–128), your students can use the temperature probe to take readings of the aquarium water over a 24-hour period. This also allows students to find the average temperature of the aquarium.

The use of measurement tools can lead to a class discussion about their advantages and limitations. Ask your students these questions:

- What are the differences between quantitative and qualitative observations?
- What are the advantages of making quantitative observations?
- How have the measurement tools used in your experiments been useful in making quantitative observations? Do you think these measurements are accurate? What are the limitations of the tools?
- Why is it advantageous to repeat experiments and average your results?

—J. S.

Scholastic pfs: File
Life Science Data Base
Physical Science Data Base
Scholastic Software, Inc.
730 Broadway
New York, New York 10003
(314) 636-8890

Available for Apple II family
$79.95 each

As a teacher trying to keep up with new developments in science, you may have been maintaining many sources of information in filing cabinets, book cases, and even closet shelves. It can be quite a chore to locate a particular resource when you need it. Technology offers you a way to organize these resources by storing information electronically in a data base. Since this information can be updated any time, the facts stored in the data base can remain current.

To use the software, students must formulate a question and then decide which field names (categories) might contain data that could answer the question. This helps them practice developing and testing a hypothesis. Also, learning to search for information in a data base and determining if the data answers the question are important skills that your students will need in the future. By using a data base, students become active learners and teachers become facilitators; your students can spend more time thinking about the topic, drawing comparisons, looking for relationships, and evaluating results.

For an example of how data base software might be used in a classroom, consider the example of an earth science unit on earthquakes. The teacher in this class will spend time discussing causes of earthquakes, locations prone to earthquake activity, and how to survive an earthquake. Rather than give a series of lectures, the teacher can have students become active participants in the learning process by asking them to find information about earthquakes in a data base. For instance, if you want students to describe the geologic characteristics of earthquake-prone regions, they may try to find out if there are common land forms and geologic rock types in these parts of the world, or how close these locations are to the plate boundaries. To obtain this information from a variety of reference books would take students a long time, but it can be obtained very quickly from an electronic data base.

The science curriculum data bases published by Scholastic contain complete teacher handbooks in addition to the facts on the disks. The strength of this material is the format presented for using and building files. Activities that can be performed with the information stored in the files are modeled, and the support materials are written with step-by-step instructions. At the beginning of each unit is an overview of goals, objectives, and the time required for each activity. Having used this material with my own students, I have found that it is an excellent way to introduce them to using a data base.

To get a better idea of how you can use a data base program in your classroom, consider this hypothetical example of a group of students who have been given an open-ended assignment to research one aspect of earthquakes using a data base. With one person acting as a recorder, the students brainstorm questions that could be answered using information stored in the data base. Their questions include:

- Why does California have so many earthquakes?
- What type of house suffers the least damage in an earthquake?
- What type of underlying geologic formation is best to build on?
- What is the relationship between injuries/deaths and types of building structures?
- Does weather play a significant role in earthquakes?
- Do earthquakes occur more often at a certain time of year?

- What is the relationship between earthquake magnitude, duration, and building damage?
- What is the relationship between the location of a fault and the damage caused by an earthquake?
- Are some faults more active than others? If so, why?

These questions are all general in nature. The data base contains information that can be used to look for trends and relationships in earthquakes around the world. It provides a collection of information to support a hypothesis, much like the observations in a science laboratory experiment.

By analyzing the questions, students then list the key points that would enable them to obtain an answer to one of their questions. If the students decide to find data to answer the question, "What type of house suffers the least damage in an earthquake?" they might use the data base to look at key points such as earthquake magnitude, duration, number of aftershocks, type of building materials, average building height, the building code, distance from a fault, underlying geologic rock type, and building damage in previous earthquakes. After finding the information in these categories, students can print the data comparing building damage in earthquakes of equal magnitude and duration to see if the local geologic structure and building types are similar. Students are then able to make conclusions based on the scientific information in the data base. A typical conclusion for this question would be that one-story wood frame homes with continuous foundations built on top of bedrock usually suffer the least damage in an earthquake. This is only one example of how data bases can be used in your classroom; the research that can be done using a data base is practically unlimited.

—J. G.

Multi-Purpose Software
In addition to using the computer as an enrichment tool for a particular science activity, you can also use the it as an evaluation tool while at the same time adding diversity and fun to your classroom. The following four programs can be used to provide additional evaluation of student progress and understanding.

Crossword Magic
Mindscape Inc.
3444 Dundee Road
Northbrook, Illinois 60062
(312) 480-7667

Available for Apple II family
$49.95

Use this program to generate crossword puzzles for your students simply by typing in a list of appropriate words on any subject. The computer interconnects the words for you, creating a crossword puzzle. Alternatively, let your students create their own puzzles about a specific topic. This requires that students learn the proper spelling of the words, and write succinct and accurate definitions as clues. The puzzles can be printed out or worked directly on the screen. I've found that nothing inspires high quality work more than the knowledge that it will be used by their peers.

—J. S.

Ten Clues
Sunburst Communications
39 Washington Avenue
Pleasantville, New York 10570
(800) 431-1934 or (914) 769-5030

Available for Apple II family
$65.00

As with *Crossword Magic*, the use of *Ten Clues* is applicable to any subject area. It is a software game based on the game Twenty Questions. A series of ten clues entices students to guess a mystery word or phrase. An option in the program allows students to select a mystery word and create their own clues. The strength of the game lies in the fact that students must differentiate between critical and variable attributes.

—J. S.

The Game Show and **Tic Tac Show**
Advanced Ideas, Inc.
2902 San Pablo Avenue
Berkeley, California 94702
(415) 526-5100

Available for Apple II family, IBM PC, and Commodore 64
$39.95 each

These two programs can also be used as evaluation tools because they allow the teacher or the student to enter clues on any subject area. *The Game Show* uses the format of a TV game show in which a host asks questions of two teams competing against each other. *Tic Tac Show* also uses a game show format based on the old standard tic-tac-toe.

—J. S.

Enrichment Software
These three programs are excellent enrichment activities that can be added to almost any science curriculum.

Rocky's Boots
The Learning Company
P. O. Box 2168
Menlo Park, California 94026
(800) 852-2255 or (415) 792-2101

Available for Apple II family, IBM PC and IBM PCjr, Tandy 1000, and Commodore 64
Color monitor required
$49.95

This program provides a great way for students to develop logical thinking and problem-solving skills while learning the basics about electronic and computer circuitry. The software features a character named Rocky the Raccoon, and it provides 40 interactive games in which students build animated machines that "boot" various objects. A package of support materials titled *Rocky's Circuits* is also available from Project SETUP, Fresno Pacific College, 1717 South Chestnut Avenue, Fresno, California 93702.

—J. S.

The Voyage of the Mimi
Holt, Rinehart & Winston
School Division
1627 Woodland Avenue
Austin, Texas 78741
(800) 782-4479

Available for Apple II family
$75.00 (*Ecosystems* software only)

A multi-media simulation that reinforces students' understanding of ecosystems. The complete program includes videotapes, student books, a teacher's guide, and four software learning modules. I have used the *Ecosystems* module in my classroom. This module focuses on endangered species and ecological balance. Students select eight species (four aquatic and four terrestrial) to inhabit an island ecosystem, with the goal of enhancing their survival. Other modules available include *Whales and Their Environment, Maps and Navigation,* and *Introduction to Computing.*

—J. S.

Oh, Deer!
MECC (Minnesota Educational Computing Corp.)
3490 Lexington Avenue North
St. Paul, Minnesota 55126
(612) 481-3500

Available for Apple II family
$45.00

A simulation in which students make decisions on how to manage a hypothetical deer population near the fictional town of Whitetail Hollow. The size of the deer herd needs to be kept in balance with the natural environment and with the level of human tolerance in the residential area. Various controversies ensue, and students explore a number of possible solutions. Finally, they develop a five-year management plan and test their decisions. *Oh, Deer!* helps students understand population dynamics, controlled and uncontrolled variables, and animal-human relationships.

—J. S.